T0269131

Materials and Sustainable Development

Materials and Sustainable Development

Michael F. Ashby

*Cambridge University and Granta Design,
Cambridge, UK*
with Didac Ferrer Balas and Jordi Segalas Coral
Universitat politecnica de Catalunya, Spain

AMSTERDAM • BOSTON • HEIDELBERG • LONDON
NEW YORK • OXFORD • PARIS • SAN DIEGO
SAN FRANCISCO • SINGAPORE • SYDNEY • TOKYO

Butterworth-Heinemann Publishing is an imprint of Elsevier

Butterworth-Heinemann is an imprint of Elsevier
The Boulevard, Langford Lane, Kidlington, Oxford OX5 1GB, UK
225 Wyman Street, Waltham, MA 02451, USA

Copyright © 2016 Elsevier Ltd. All rights reserved.

No part of this publication may be reproduced or transmitted in any form or by
any means, electronic or mechanical, including photocopying, recording, or any
information storage and retrieval system, without permission in writing from the
publisher. Details on how to seek permission, further information about the Publisher's
permissions policies and our arrangement with organizations such as the Copyright
Clearance Center and the Copyright Licensing Agency, can be found at our website:
www.elsevier.com/permissions

This book and the individual contributions contained in it are protected under
copyright by the Publisher (other than as may be noted herein).

Notices
Knowledge and best practice in this field are constantly changing. As new research and
experience broaden our understanding, changes in research methods, professional
practices, or medical treatment may become necessary.

Practitioners and researchers must always rely on their own experience and knowledge
in evaluating and using any information, methods, compounds, or experiments
described herein. In using such information or methods they should be mindful
of their own safety and the safety of others, including parties for whom they have a
professional responsibility.

To the fullest extent of the law, neither the Publisher nor the authors, contributors, or
editors, assume any liability for any injury and/or damage to persons or property as a
matter of products liability, negligence or otherwise, or from any use or operation of
any methods, products, instructions, or ideas contained in the material herein.

British Library Cataloguing in Publication Data
A catalogue record for this book is available from the British Library

Library of Congress Cataloging-in-Publication Data
A catalog record for this book is available from the Library of Congress

For information on all Butterworth-Heinemann Publishing
visit our website at http://store.elsevier.com

ISBN: 978-0-08-100176-9

Working together
to grow libraries in
developing countries

ELSEVIER   Book Aid International

www.elsevier.com • www.bookaid.org

Contents

ACKNOWLEDGEMENTS xi
PREFACE...xiii

CHAPTER 1 **Background: Materials, Energy and Sustainability** 1

    1.1 Introduction and Synopsis 2
    1.2 Sustainable Development – A Brief History 3
    1.3 Materials – An Even Briefer History 7
    1.4 Critical Materials 11
    1.5 Energy – Units and Quantities 16
    1.6 Resources, Consumption, Population, Affluence and Impact 17
    1.7 Summary and Conclusions 20
    1.8 Exercises 20

CHAPTER 2 **What is a "Sustainable Development"?** 27

    2.1 Introduction and Synopsis 28
    2.2 What Does "Sustainability" Mean? 28
    2.3 Defining "Sustainable Development" 30
    2.4 Articulations of Sustainable Development 33
    2.5 Summary and Conclusions 36
    2.6 Exercises 37

CHAPTER 3 **Assessing Sustainable Developments: The Steps** 39

    3.1 Introduction and Synopsis 39
    3.2 Dealing with Complex Systems 40
    3.3 A Layered Approach to Assessing a Sustainable Development...42
    3.4 Assembling the Layers 50
    3.5 Summary and Conclusions 51
    3.6 Exercises 51

CHAPTER 4 **Tools, Prompts and Check-Lists** 55

    4.1 Introduction and Synopsis 56
    4.2 Step 1: Clarifying the Prime Objective......................56

vi   **Contents**

4.3 Step 2: Stakeholder Analysis 57
4.4 Step 3: Fact-Finding 60
4.5 Step 4: Informed Synthesis 66
4.6 Step 5: Reflection on Alternatives 71
4.7 Summary and Conclusions 71
4.8 Appendix: Creativity Aids – A Brief Survey 72
4.9 Exercises 82

CHAPTER 5   **Materials Supply-Chain Risk 85**

5.1 Introduction and Synopsis 85
5.2 Emerging Constraint on Material Sourcing and
         Usage 86
5.3 Price Volatility Risk..88
5.4 Monopoly of Supply and Geopolitical Risk 89
5.5 Conflict Risk 91
5.6 Legislation and Regulation Risk 92
5.7 Abundance Risk...94
5.8 Changing Expectation of Corporate Responsibility....95
5.9 Managing Risk 96
5.10 Summary and Conclusions 97
5.11 Exercises 97

CHAPTER 6   **Corporate Sustainability and Materials 101**

6.1 Introduction and Synopsis 101
6.2 Corporate Social Responsibility and Sustainability
         Reporting 102
6.3 Case Studies: Corporate SRs 105
6.4 Summary and Conclusions 108
6.5 Exercises 108

CHAPTER 7   **Introduction to Case Studies 111**

7.1 Introduction and Synopsis 111
7.2 The Structure of the Case Studies 112
7.3 Articulations of Sustainable Development
         That Went Wrong 113
7.4 Summary and Conclusions 115
7.5 Exercises 116

CHAPTER 8   **Scaling Up Biopolymer Production.................... 117**

8.1 Introduction and Background Information...............118
         8.1.1 Background Information 119
8.2 Prime Objective and Scale 120

8.3 Stakeholders and Their Concerns 120
8.4 Fact-Finding...122
8.5 Synthesis with the Three Capitals 127
    8.5.1 Natural Capital 129
    8.5.2 Manufactured and Financial Capital 129
    8.5.3 Human and Social Capital 130
8.6 Reflection on Alternatives................................130
    8.6.1 Short Term 130
    8.6.2 Longer Term 131
8.7 Related Projects 132

CHAPTER 9 Wind Farms... 135

9.1 Introduction and Background 135
    9.1.1 Background Information 136
9.2 Prime Objective and Scale................................138
9.3 Stakeholders and Their Concerns 138
9.4 Fact-Finding...140
9.5 Synthesis with the Three Capitals 145
9.6 Reflection on Alternatives................................147
9.7 Related Projects 148

CHAPTER 10 Case Study: Electric Cars................................ 151

10.1 Introduction and Background 151
    10.1.1 Background Information...............................152
10.2 Prime Objective and Scale 153
10.3 Stakeholders and Their Concerns........................153
10.4 Fact-Finding 155
10.5 Synthesis with the Three Capitals 161
10.6 Reflection on Alternatives 163
    10.6.1 Short Term 163
    10.6.2 Long Term 164
10.7 Related Projects 165

CHAPTER 11 Lighting 167

11.1 Introduction and Background Information 167
    11.1.1 Background Information...............................168
11.2 Prime Objective and Scale 169
11.3 Stakeholders and Their Concerns........................170
11.4 Fact-Finding 172
11.5 Synthesis with the Three Capitals 177
11.6 Reflection on Alternatives 179
11.7 Suggested Projects 179

CHAPTER 12 Solar PV **181**

    12.1 Introduction and Background Information 182
        12.1.1 Background Information...........................182
    12.2 Prime Objective and Scale 183
    12.3 Stakeholders and Their Concerns........................184
    12.4 Fact-Finding 186
    12.5 Synthesis with the Three Capitals 191
    12.6 Reflection on Alternatives 192
    12.7 Suggested Projects 194

CHAPTER 13 Bamboo for Sustainable Flooring **197**

    13.1 Introduction and Background Information 198
        13.1.1 Background Information...........................199
    13.2 Prime Objective and Scale 200
    13.3 Stakeholders and Their Concerns........................200
    13.4 Fact-Finding 202
    13.5 Synthesis with the Three Capitals 207
    13.6 Reflection on Alternatives 208
    13.7 Suggestions for Related Projects.........................210

CHAPTER 14 The Vision: A Circular Materials Economy **211**

    14.1 Introduction and Synopsis 212
    14.2 The Ecological Metaphor.....................................213
    14.3 The Scale of the Vision..217
    14.4 The Circular Materials Economy 219
    14.5 Creating a Circular Materials Economy.................222
        14.5.1 Better Stuff: Improved Materials
                Technology 222
        14.5.2 Better Design 223
        14.5.3 Better Business Models 231
        14.5.4 Better Behavior: Regulation, Social
                Adaptation and Change of Life-Style 233
    14.6 Summary and Conclusions 234
    14.7 Exercises 236

CHAPTER 15 Data, Charts and Databases **241**

    15.1 Introduction and Synopsis 242
    15.2 The CES Sustainability Database 242
    15.3 Using the Elements Data-Table 244

15.4 Using the Materials Data-Table 246
15.5 Using the Power Systems Data-Table....................247
15.6 Using the Energy Storage Systems Data-Table.......249
15.7 Using the Legislation and Regulations
Data-Table..250
15.8 Using the Nations of the World Data-Table 252
15.9 Summary and Conclusions 257

CHAPTER 16 Guidance for Instructors.................................... **259**

16.1 Introduction and Synopsis 260
16.2 Problem-Based Learning.....................................260
16.3 PBL and Sustainable Development.....................261
16.4 Organizing the Project; Scheduling the
Activities 262
16.5 Assessment 266
16.6 Feedback from Students (UPC Course
"Sustainable Design" 2012, 2013)......................267
16.7 Summary and Conclusions 270
16.8 Suggestions for Further Projects.........................271

Appendix: Useful Numbers **275**
Index **303**

Acknowledgements

We wish to acknowledge the insights and helpful critical reviews of a number of colleagues, among them Professor Julian Allwood, Dr Jon Cullen, Professor David Cebon, Ethan Attwood and Jennifer Bruce of Cambridge University, Professor Karel Mulder of the Technical University of Delft, Professor Peter Goodhew of Liverpool University, and Professor Alexander Wanner of the Karlsruhe Institute of Technology. Special thanks are due to Professor John Abelson and the students of Class ENG 571 at the University of Illinois in Urbana and those of the Masters in Sustainability Program at the Universitat politecnica de Catalunya in Barcelona who trialed the methods described in the book and commented in detail on the manuscript. We also wish to thank the reviewers of an earlier draft of the manuscript for their helpful suggestions and comments, particularly Dr Deborah Andrews of London South Bank University, Dr Magdalena Titirici of Queen Mary College, London, Professor Jeffrey Fergus of Auburn University, Alabama, Professor Trevor Harding of California Polytechnic State University, California, Professor Victor Li of the University of Michigan, and Daniel Kenning of Splendid Engineering, Chelmsford, UK. Finally, we wish to acknowledge the involvement of Dr Tatiana Vakhitova, Dr Claes Fredriksson, Dr James Goddin, Kim Marshall and Hannah Melia and our many other associates at Granta Design, Cambridge, without whom this study would not have been possible.

Michael F. Ashby, Didac Ferrer Balas and Jordi Segalas Coral
Cambridge, 2014

Preface

The last two centuries have been an era of material plenty. Over much of this time our ability to locate, mine, refine and use materials increased and, with it, material availability and affordability. The dominant business model, in material terms, was one of *take – make – use – dispose*, with focus on economic growth through increasing consumption. Material supply was not (except in wartime) a major issue. Trade tended to be national rather than global. There was relatively little control over the way materials were used or what happened to them at the end of product life. Material prices, in real terms, were static or falling.

Today, the picture looks rather different. Global population now exceeds seven billion and continues to rise. More significantly, global affluence is also rising and most rapidly so in the most populous nations. It is hard to ignore the evidence of the limited capacity of the eco-system to cope with the demands we now place on it. Some are environmental. Air pollution in cities, water pollution of rivers, chemical contamination of land should, you would think, be under our control, yet we often fail to contain them. The increasing complexity of products has created a dependence on a larger number of elements, some comparatively rare. These are sourced globally and used to make products that are traded on a global scale. The era of "easy oil" is approaching an end and we are now prepared to drill through kilometres of rock beneath a kilometre of water to find it. Manufacturing nations increasingly compete for exclusive rights to mineral resources world-wide in order to safeguard their industrial capacity. Nations from which these resources were, in the past, freely traded now exert more control over supply to protect domestic consumption or for geo-economic or political purposes. New and expanding legislation controls many aspects of manufacturer responsibility, product design, material usage and material disposal. The public, stakeholders and government increasingly judge corporate success not just in financial terms but in terms of stewardship of the environment and welfare of its workforce and that of the local economy of the communities in which it operates.

Sustainable development means living (and thriving) in this changing environment. We cannot live without using energy, water and materials. The challenge is to use the so as "to meet them needs of today without compromising the ability of future generations to meet their own needs[1]" and to do so in ways that are equitable and within the rule of law.

This book provides a structure and framework for analysing sustainable development and the role of materials in it. Chapter 1 sets the scene. Chapters 2 to 6 describe the method, first in a simple, broad way, then with added details and aids. The method is illustrated in a set of six case studies, introduced in Chapter 7 and presented in Chapters 8 to 13. Chapter 14 takes a broader view leading up to the idea of a circular materials economy. Chapter 15 describes a database that is designed to support the analysis of sustainable developments, and illustrates its content and use. Chapter 16 is aimed at instructors. It summarizes our experience in using the approach developed in the book and suggests the way the approach can be used for project or activity-based teaching. The book ends with an Appendix that compiles useful numbers.

The aim is to introduce ways of exploring sustainable development to students of materials science and engineering at the bachelors and masters levels. It recognizes the complexity inherent in discussions of sustainability and shows how to deal with it in a systematic way. There is no completely "right" answer to questions of sustainable development – instead, there is a thoughtful, well-researched response that recognizes the conflicting priorities of the environmental, the economic, the legal and the social aspects of a technological change. The intent is not to offer solutions but rather to improve the quality of discussion and enable informed, balanced debate.

---

[1]The Brundtland Report of the World Council on Economic Development (WCED, 1987).

CHAPTER 1

Background: Materials, Energy and Sustainability

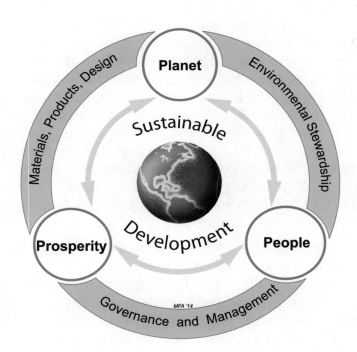

Chapter Outline

1.1 Introduction and Synopsis   2
1.2 Sustainable Development – A Brief History   3
1.3 Materials – An Even Briefer History   7
1.4 Critical Materials   11
1.5 Energy – Units and Quantities   16

1.6 Resources, Consumption, Population, Affluence and Impact   17
1.7 Summary and Conclusions   20
1.8 Exercises   20
Further Reading   23

Materials and Sustainable Development. http://dx.doi.org/10.1016/B978-0-08-100176-9.00001-3
Copyright © 2016 Elsevier Ltd. All rights reserved.

1.1 INTRODUCTION AND SYNOPSIS

Sustainable development has to do with our relationship with the natural environment on which we depend for food, water, energy and raw materials. But there is much more to it than that. It is also about our relationship with the global economic system in which we source raw materials, manufacture products and trade. And perhaps most important, it has to do with our relationship with each other, meaning the values of the society in which we live and its relationship with other societies. Understandably, not everyone perceives sustainable development in the same way. An environmentalist might judge it by its contribution to the protection and nurture of the natural environment, preserving clean air, pure water, productive land and a thriving biosystem. A humanist might instead look for its contribution to the generation and sharing of knowledge and understanding. The corporate view of sustainable development might accept these but see the financial health of the corporation as the ultimate metric. Thus even the simplest of sustainable developments has at least three facets: environmental, social and economic.

The Western view of development is one of economic growth based on urbanization and technical innovation, using global markets to source resources and to distribute goods and services. Western democracies have become good at this, harnessing the energy and resources of the natural environment to provide goods and services. Its proponents recognize that natural resources are finite in extent but point out that technical advance has, in the past, more than offset the depletion of resources and that there is no reason to think that it will fail to do so in the future. The economic histories of developed nations suggest a natural progression from early agrarian society through industrialization to a postindustrial economy in which wealth increases faster than population thereby enabling further economic growth, and with it, an evolving society. In this view, less-developed nations should try to emulate the Western model, opening their countries to Western values and to global trade in resources, goods and services.

The concept of *sustainable* development challenges this techno-centric view. The view of nature as a resource to be tapped to meet present-day human needs ignores the needs both of other forms of life and of future generations of humans. The focus on disposable income, with Gross Domestic Product (GDP) seen as a metric of national prosperity, prioritizes wealth and owner-ship of goods above quality of life and individual self-attain-ment above the common good. Recent history, particularly, has prompted self-analysis in the West: the banking crisis of 2008 and the failure of democracy to root itself immediately in the Arab spring raise questions about the Western ideal of free-mar-ket economies. Most telling, perhaps, is the simple fact that it is not possible for all nations to replicate the lifestyle of the West – the planet's atmosphere is not able to absorb the resulting emissions, its natural hydrological cycle cannot provide the nec-essary freshwater, and the present-day international scramble to secure access to mineral resources suggests that these, too, are under pressure.

Well, there is a lot here. How can we get to grips with it all? The best place to start is the big picture. What is the background to current thinking about sustainable development and how is it evolving? How has our dependence on materials arisen? And where does the energy needed to extract and process them come from? We start with brief histories of all three.

1.2 SUSTAINABLE DEVELOPMENT – A BRIEF HISTORY

Technological development without regard for environmental and social impacts brings undesired consequences: degradation of air, water and land, loss of biodiversity, resource depletion, and increasing inequality. This realization may seem recent, but it is not new. Thomas Malthus, writing in 1798, foresaw the link between population growth and resource depletion, predict-ing gloomily that the demands of a growing population would, sooner or later, outstrip the capacity of the earth to support it. Almost 200 years later, a group of scientists known as the Club of Rome reported their modeling of the interaction of population growth, resource depletion and pollution, concluding that "*if (current trends) continue unchanged …. humanity is destined to reach the natural limits of development within the next 100 years*" (Meadows et al., 1972). The report generated both consternation and criticism, largely on the grounds that the modeling did not allow for scientific and technological advance. But in the last decade thinking about this broad issue has reawakened. There is a growing acceptance that, in the words of another distinguished report, *"many aspects of developed societies are approaching...saturation, in the sense that things cannot go on growing much longer without reaching fundamental limits. This does not mean that growth will stop in the next decade, but that a declining rate of growth is foreseeable in the lifetime of many people now alive. In a society accustomed...to 300 years of growth, this is something quite new, and it will require considerable adjustment"* (WCED, 1987).

Table 1.1 lists nine documents that have had profound influence on current thinking about the effects of human activity on the environment. The publications span a little over 50 years. The starting point for today's thinking about sustainable development is the report of the Brundtland Commission of 1987 (WCED, 1987), Figure 1.1. It makes the observation that *"- each community, each country, strives for survival and prosperity with little regard for the impact on others. Some consume the earth's resources at a rate that would leave little for future generations. Others ...consume too little and live with the prospects of hunger, squalor, disease and early death."* The report formulates an ideal: *"Sustainable development is development that meets the needs of the present without compromising the ability of future generations to meet their own needs"*.

A series of initiatives since the Brundtland report have created the wide-ranging interpretation of sustainable development that we see today. The Montreal Protocol (UNEP, 1989) banned substances that damage the ozone layer and has largely fulfilled its aims. The Rio Earth Summit (UNEP, 1992), endorsed by 180 nations, set out 27 principles supporting sustainable development. After 10 years, the World Summit on Sustainable Development (WSSD, 2002), meeting in Johannesburg, reaffirmed the commitment of 183 countries to achieve sustainable development objectives, among them are as follows:

- To halve the number of people in poverty by 2015.
- To halve the proportion of people without access to clean drinking water by 2015.

Table 1.1 Landmark Publications	
Date, Author and Title	**Subject**
1962 Rachel Carson, "Silent Spring" (Carson, R., 1962)	Meticulous examination of the consequences of the use of the pesticide DDT and the impact of technology on the environment.
1972 Club of Rome, "Limits to Growth" (Meadows et al., 1972)	The report that triggered the first of a sequence of debates in the twentieth century on the ultimate limits imposed by resource depletion.
1972 The Earth Summit in Stockholm (UNEP, 1972)	The first conference convened by the United Nations, which resulted in a declaration on the state of the global environment and 26 principles for protecting it.
1987 The UN World Commission on Environment and Development "Our common future" (WCED, 1987)	Known as the Brundtland Report, it defined the principle of sustainability as "Development that meets the needs of today without compromising the ability of future generations to meet their own needs".
1987 Montreal Protocol (UNEP, 1989)	The International Protocol to phase out the use of chemicals that deplete ozone in the stratosphere.
1988 Formation and first meeting of the IPCC	The Intergovernmental Panel on Climate Change (IPCC) was set up by UNEP and WHO to provide the world with a clear scientific view on climate change and its likely consequences.
1992 Rio Declaration (UNEP, 1992)	An International statement of the principles of sustainability, building on those of the 1972 Stockholm Earth Summit.
1998 Kyoto Protocol (UNFCC, 1997)	An international treaty to reduce the emissions of gases that, through the greenhouse effect, cause climate change.
2001 Stockholm Convention (UNIDO, 2001)	The first of a series of meetings to agree an agenda for the control and phase-out of Persistent Organic Pollutants
2002 Johannesburg World Summit (WSSD, 2002)	The Summit, held 10 years after the Earth Summit in Rio, identified five priorities for Sustainable Development: health, energy, water, agriculture and biodiversity.
2014 IPCC Fifth Assessment Report (IPCC, 2014)	The Report provides an update of knowledge on the scientific, technical and socioeconomic aspects of climate change. It puts greater emphasis on assessing the socioeconomic consequences and their implications for sustainable development.

- To use chemicals in ways that minimize adverse effects on human health and the environment.
- To diversify energy supply, increase provision of energy from renewable sources and accelerate the development of energy-efficient technologies.

The concept of sustainable development has gained acceptance within the business community. Many forward-looking companies and businesses now integrate sustainability into corporate thinking and practice. The World Business Council for Sustainable Development, a coalition of 165 of the world's largest companies express *"a shared commitment to sustainable development via the three pillars of economic growth, ecological balance and social progress"*.

FIGURE 1.1
The cover of the Brundtland commission report, the starting point for discussing sustainable development.

All these are statements of intent – ideals, if you like. They feel right; few people would disagree with them. But they leave four questions unanswered:

- How do we achieve sustainable development?
- How do we measure progress in achieving it?
- What does it mean in engineering practice?
- Given the title of this book, how do materials fit in?

These questions provide a lead-in to the next three chapters, in which we attempt to answer them. But first, a short digression to introduce the history of materials.

1.3   MATERIALS – AN EVEN BRIEFER HISTORY

Materials have a long history (Figure 1.2). The earliest "engineering" materials were those of nature: stone, wood, fibers, bark, skin, hide and bone. Their applications were largely mechanical (housing and tools) or thermal (clothing and protection). Subsequent advances in technology and science-stimulated bursts of material development: thermochemistry in the seventeenth century, then of electrochemistry in the eighteenth and nineteenth centuries, so that by the mid-twentieth century almost the entire periodic table was accessible for engineering purposes. The growth of polymer and ceramic sciences in the first half of the twentieth century delivered new structural materials. They were followed, in the second half of the century, by the development of materials with greater functionality: piezoelectrics, thermoelectrics, magnetic, and above all semiconducting and optoelectronic materials. The time steps of Figure 1.2 are nonlinear, with big time steps at the bottom, small at the top, reflecting the accelerating rate of materials innovation. It shows no sign of slowing. Since 1971, when the last of the actinides was isolated, material developers have had access to all 92 of the stable elements of the Periodic Table.

In a surprisingly short space of time, we have become dependent on this treasure chest of elements and the materials made from them. Even half a century ago, engineering relied on relatively few, widely available elements (Table 1.2). Many products, today, draw on a much wider spectrum, some of which occur only as lean, highly localized ores, or are extracted only as a by-product of another element, making their supply uncertain. Our dependence on some of

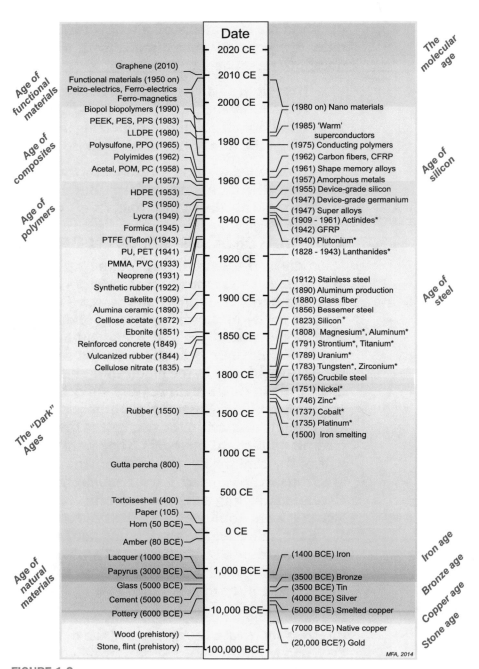

FIGURE 1.2

A materials time-line. The timescale is nonlinear with big steps at the bottom and small at the top. The figure brings out the accelerating rate of material development and, with it, our dependence on an ever larger number of them.

Table 1.2	The Increasing Diversity of Elements Used in Materials and Devices over the Past 75 Years	
Alloys and Devices	**Changing Demand for Elements over Time**	
	75 Years Ago	**Today**
Iron-based alloys*	Fe, C	Al, Co, Cr, Fe, Mn, Mo, Nb, Ni, Si, Ta, Ti, V, W
Aluminum alloys*	Al, Cu, Si	Al, Be, Ce, Cr, Cu, Fe, Li, Mg, Mn, Si, I, V, Zn, Zr
Nickel alloys*	Ni, Cr	Al, B, Be, C, Co, Cr, Cu, Fe, Mo, Ni, Si, Ta, Ti, W, Zr
Copper alloys*	Cu, Sn, Zn	Al, Be, Cd, Co, Cu, Fe, Mn, Nb, P, Pb, Si, Sn, Zn
Magnetic materials*	Fe, Ni, Si	Al, B, Co, Cr, Cu, Dy, Fe, Nd, Ni, Pt, Si, Sm, V, W
Displays	W	Eu, Ge, Ne, Si, Tb, Xe, Y
(Micro) electronics	Cu, Fe, W	As, Ga, In, Sb, Si
Low-C energy (Solar, Wind)	Cu, Fe	Ag, Dy, Ga, Ge, In, Li, Nd, Pd, Pt, Re, Se, Si, Sm, Te, Y

*Data from the composition fields of records in the CES EduPack '14 Level 3 database, Granta Design, (2014).

these is now so extreme that Governments classify them as "critical" and regard access to them as a strategic necessity.

And we use them in enormous quantities. Speaking globally, we consume roughly 10 billion (10^{10}) tonnes of engineering materials per year, an average of 1.4 tonnes per person. Figure 1.3 gives a perspective: it is a bar chart of the primary production of the materials used in the greatest quantities. It has some interesting messages. On the extreme left, for calibration, are hydrocarbon fuels – oil and coal – of which we currently consume a colossal 9 billion (9×10^9) tonnes per year. Next, moving to the right, are metals. The scale is logarithmic, making it appear that the production of steel (the first metal) is only a little greater than that of aluminum (the next); in reality, the production of steel exceeds, by a factor of ten that of all other metals combined. Steel may lack the high-tech image that attaches to materials like titanium, carbon-fiber reinforced composites and (most recently) nanomaterials, but make no mistake, its versatility, strength, toughness, low cost and wide availability

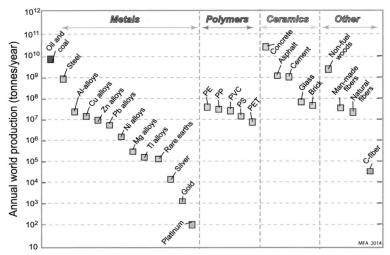

FIGURE 1.3
The annual world production of some of the materials on which industrialized society depends. The scale is logarithmic.

are unmatched. At the other extreme are the platinum-group metals. Their quantities are small, but their importance is large.

Polymers come next. Before 50 years, their production was tiny. Today the production, in tonnes per year, of the five commodity polymers; polyethylene (PE), polypropylene (PP), polyvinyl chloride (PVC), polystyrene (PS) and polyethylene-terephthalate, (PET) is comparable with that of aluminum; if measured in volume per year they approach steel. The really big ones, though, are the materials of the construction industry. Steel is one of these, but the production of wood for construction purposes exceeds even that of steel when measured in tonnes per year (as in the diagram), and since it is a factor of 10 lighter, wood totally eclipses steel when measured in volume per year. Bigger still is the production of concrete, which exceeds that of all other materials. The other big ones are asphalt (roads), cement (most of which goes into concrete), brick and glass.

Fibers, too, are produced in very great quantities. Natural fibers have played a role in human life for tens of thousands of years, and continue to do so. Even larger, today, is the production of man-made fibers for both textiles and for industrial use. It includes today's production of carbon fiber. Just 30 years ago, this material would not have crept onto the bottom of the chart. Today its production is approaching that of titanium and is growing much more quickly.

Table 1.3 Critical Materials–US List
Antimony
Beryllium
Bismuth
Cerium
Chromium
Cobalt
Dysprosium
Erbium
Europium
Gallium
Germanium
Indium
Iridium
Lanthanum
Lithium
Manganese
Neodymium
Osmium
Palladium
Platinum
Praseodymium
Rhodium
Ruthenium
Samarium
Scandium
Tantalum
Tellurium
Terbium
Thulium
Tin
Tungsten
Yttrium

1.4   CRITICAL MATERIALS

Materials today are sourced and traded globally. The ores and feed-stock needed to make some of them are widely available, allowing a free market to operate smoothly. Others, however, are geographically more restricted; the nations from which they come may limit supply for economic or political reasons; and political instability or conflict there or in neighboring states may interrupt supply completely. The supply-chain of cobalt, for instance, was disrupted recently by war in the Democratic Republic of the Congo, and that of rare-earth metals has not operated smoothly in the recent past because of monopoly constraints that led to shortages and unstable pricing.

Materials are classified as "critical" if access to them could be limited or they are essential for national security or important economically. Governments draw up lists of these critical materials and develop strategies to ensure supply by negotiating exclusive supply agreements, seeking new sources, or stockpiling. The importance attached to critical materials is evidenced by the sheer number of such studies – at least five have appeared in the last three years, among them those of the US Geological Survey[1], the US Department of Energy[2], the American Physical Society[3], the European Union[4] and the British Geological Survey[5]. The lists are not identical because "criticality" depends to some extent on the needs and domestic resources of the nation, but they have a great deal in common. Those that appear in the US lists are shown in Table 1.3. The lists act as warnings, alerting manufacturers of potential supply risk.

Figure 1.4 shows a specific example. Lamp bulbs in the last century had tungsten filaments – tungsten is a critical material – but the rest of the device required only glass and widely available metals. The high-efficiency lamp bulb of today still contains tungsten but depends also on a cocktail of rare-earth phosphors and inert gases to achieve the desired spectrum of light[6]. The critical metals are colored red.

---

[1]US Geological Survey (2009) (http://minerals.usgs.gov/east/critical/index.html).
[2]US Department of Energy (2010) (www.energy.gov/news/documents/criticalmaterialsstrategy.pdf).
[3]American Physical Society (2010) (http://www.aps.org/units/fps/newsletters/201107/jaffe.cfm).
[4]European Union (2010) http://ec.europa.eu/enterprise/policies/raw-materials/critical/.
[5]British Geological Survey (2012) (www.bgs.ac.uk/downloads/start.cfm?id=2643).
[6]Elements in lamps US Department of Energy www.energy.gov/news/documents/criticalmaterialsstrategy.pdf.

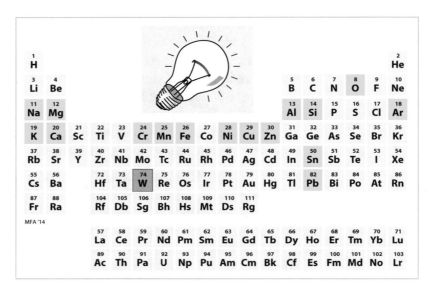

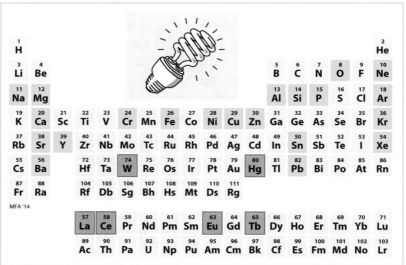

FIGURE 1.4
The materials of the two lamps mapped onto the periodic table. Critical materials are colored red, other materials used in the product are in darker yellow.

Figure 1.5 is a still more striking example. The phone of the 1960s contained just nine metals for the frame, conductors, magnets, springs and contacts, plus the carbon, hydrogen and nitrogen of the plastic casing. Mobile communication devices of today contain at least fifty three elements[7] of which twenty one, again colored red in the figure, are listed as "critical". The same is true of almost all consumer electronics, of communication systems and transport and (above all) of defence and security-related equipment.

Figure 1.6 shows a third example: a comparison of the materials in aircraft engines. The early engine used few materials, of which only tungsten (spark plugs) was in any way critical. The gas turbine of today depends on many more, and we are speaking not of milligrams but of kilograms. The message is clear: products on which our current way of life depends are, in material terms, far more complex than those of 50 years ago. This dependency bring with it risk. We examine this risk and ways of dealing with it in more detail in Chapter 5.

Producing Nation	Platinum Tonnes/year, 2011
South Africa	139,000
Russia	26,000
Canada	10,000
Zimbabwe	9,400
United States	3,700
Colombia	1,000
Other countries	2,500
World	**192,000**

Example. Platinum performs a unique function as a catalyst for chemical reactions. It has Critical Materials status in both US and European lists. From which nations is platinum sourced? Why is it listed as critical?

Answer. The United States Geological Survey publishes annual tables[8] of world mineral production, listing the countries of origin. The adjacent table shows its data for platinum. Over 72% derives from a single nation, South Africa. This is an example of supply-chain concentration, a cause for concern for users of platinum. Figure 1.3 shows that platinum is one of the least abundant of the elements. These facts, plus its unique properties as a catalyst, account for its critical status.

---

[7]Elements in i-phone from Sheffield Hallam University study, cited by http://www.greatrecovery.org.uk.
[8]http://minerals.usgs.gov/minerals/pubs/commodity/platinum/myb1-2012-plati.pdf.

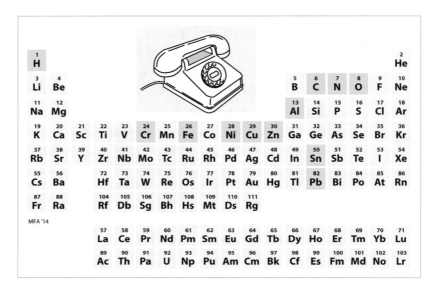

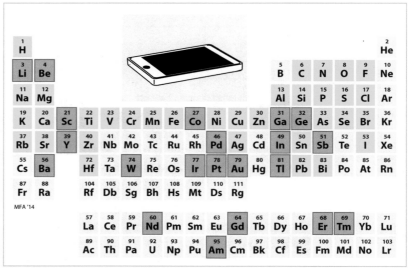

FIGURE 1.5
The elements in an electrical device of the 1950s and those in a present-day phone mapped onto the periodic table. Critical materials are colored red, other materials used in the product are in darker yellow.

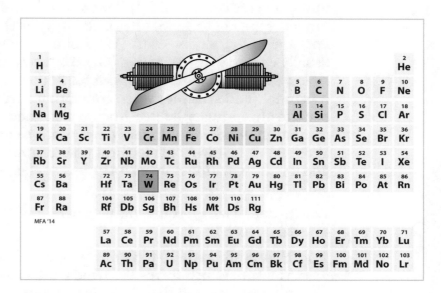

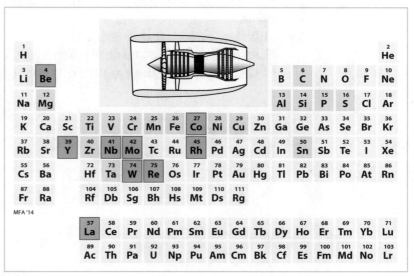

FIGURE 1.6
The elements in an early aircraft engine and a gas turbine of today mapped onto the periodic table.
Critical materials are colored red, other materials used in the product are in darker yellow.

1.5 ENERGY – UNITS AND QUANTITIES

Issues of sustainability are inseparable from those of energy. The SI unit of energy is the Joule (J) but because a Joule is very small, we generally use kJ (10^3J), MJ (10^6J), GJ (10^9J) or TJ (10^{12}J) as the unit. Power is Joules/sec, or Watts (W), but a Watt, too, is small so we usually end up with kW, MW or GW. The everyday unit of electrical energy is the kWh, one kW drawn for 3600 s, so 1 kWh = 3.6 MJ. It is usual (and sensible) when comparing energies to relate them back to the primary source from which they are drawn – most commonly, fossil fuel. Modern power stations have an efficiency of about 36%, so 1 kWh of electrical power drawn from the national grid of most nations, means a primary energy consumption of 3.6/0.36 = 10 MJ.

Where does energy come from? There are ultimately just four sources (Figure 1.7):

- Hydrocarbon fuels, the sun's energy in fossilized form.
- The sun, which drives the winds, wave, hydro, photoelectric phenomena and the photochemical processes that give biomass.
- The moon, which drives the tides.
- Nuclear decay of unstable elements inherited from the creation of the earth, providing geothermal heat and nuclear power.

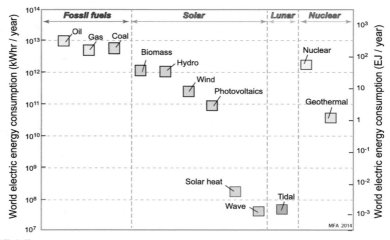

FIGURE 1.7
The annual world consumption of energy by source. The units on the left are kWh, those on the right are exajoules (10^{18} J).

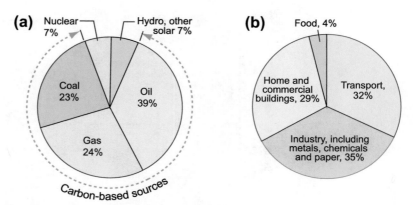

FIGURE 1.8
World energy (a) Production, by source and (b) Consumption, by use. The nonrenewable carbon-based fuels oil, gas and coal account for 86% of the total.

All four are ultimately finite but the timescale for the exhaustion of the last three is so large that it is safe to regard them as infinite.

How much energy do we use in a year? When speaking of world consumption the unit of convenience is the exajoule, symbol EJ, a billion billion (10^{18}) joules. The value today (2014) is about 590 EJ/year and, of course, it is rising. Figure 1.8(a) shows where it comes from. Fossil fuels dominate the picture, providing about 86% of the total. Nuclear delivers about 7%. Renewables – hydro, wind, wave, biomass, solar heat and photovoltaics – add up to just another 7%. These sun-driven energy pools are enormous, but unlike fossil and nuclear fuels, which are concentrated, they are distributed, making the energy hard to capture.

Where does the energy go? Most of it into three big sectors: transport, buildings (heating, cooling, lighting) and industry (Figure 1.8(b)). Making materials is a big part of the "Industrial" sector, consuming about 21% of global energy.

1.6   RESOURCES, CONSUMPTION, POPULATION, AFFLUENCE AND IMPACT

The planet's population, now over 7 billion, is continuing to grow. The present growth rate is 1.3% per year, causing the population to double every 55 years. This growth, by itself, is a cause for concern – we already consume resources at a faster rate than nature can replace them. Seeing this simply in terms of population, however, is

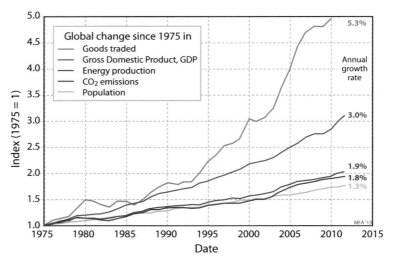

FIGURE 1.9
Growth of global trade in goods, gross domestic product (GDP), energy use, CO_2 emissions, and global population, 1975–2012. (World Bank Development Indicators, 2012).

to underestimate the problem. Over the last 40 years, growth in global trade, GDP, energy consumption and carbon emissions have all grown faster than the population (Figure 1.9). People are not only getting more numerous, they are also getting more affluent.

The *IPAT* equation[9] is an attempt to describe how growing population, affluence, and technology impact the environment. It makes the assumption that Human Impact (*I*) on the environment equals the product of Population, *P*, Affluence, *A* and Technology, *T*:

$$I = P \times A \times T.$$

Here *P* is the population of a region – it could be a single nation or it could be the whole world. Affluence, *A*, describes the spending power and, with it, the resource consumption and emissions of an average member of the population. A commonly used proxy for affluence is the GDP per capita, though it is not a very good one. The variable *T* is the resource intensity of the production of affluence, that is; the resources consumed in creating, transporting, using and disposing of the goods and services that create one unit of GDP. Improvements in efficiency can reduce resource intensity, reducing *T*.

---

[9]Much more can be said about the IPAT equation. For a fuller discussion see https://notendur. hi.is//~bdavids/UAU_102/Readings/Chertow.pdf.

Since technology can affect environmental impact in more than one way, the unit for T is chosen for the situation under study. For example, if human impact on climate change is being measured, an appropriate unit for T might be greenhouse gas emissions per unit of GDP.

Differentiating the IPAT equation we find

$$\frac{dI}{I} = \left(\frac{dP}{P}\right)_{A,T} + \left(\frac{dA}{A}\right)_{P,T} + \left(\frac{dT}{T}\right)_{P,A}.$$

The global population P is growing at 1.3% per year and the growth of affluence A, measured by that of global GDP, has averaged 3% per year over the last decade, so for impact I to remain static, technology T must bring an annual decreased of 4.3% in resource intensity per year.

The IPAT equation gives qualitative insight, but to make it quantitative it has to be refined. The parameters P, A, and T are not independent (as the equation assumes) so it is more accurate to rewrite the equation as

$$I = f(P, A, T).$$

As an example, rising affluence slows population growth; wealthy people have fewer children. As a second example, doubling technological efficiency reduces T by 50%, but experience shows that it does not reduce environmental impact I by the same factor. This is because efficiency drives down prices, stimulating additional consumption, a phenomenon known as the *rebound effect*. Beyond that, increased efficiency in the use of energy and materials stimulates economic growth, resulting in a net increase in resource consumption and associated emissions.

So we are in an exceptional period. Global population is increasing – it has done that throughout most of the world's history. But affluence is also increasing, particularly in the world's most populous nations, and at a faster rate than at any time in history. The combination is putting natural resources, national economies, the environment and social structures under unprecedented pressure. Adapting our collective behaviour to deal with these pressures is at the core of sustainable development.

1.7 SUMMARY AND CONCLUSIONS

Sustainable development is a *systems* problem. Visionary individuals (Malthus, Rachel Carsons, Meadows – see Table 1.1) perceived both this and the risks it implies, but it was not until the 1980s that the importance of thinking in holistic terms took hold. Since then numerous studies, most recently those of the Intergovernmental Panel on Climate Change (IPCC), have highlighted the potential problems for the future inherent in the way we live at present.

Materials are an important part of this system. Recent technological developments, particularly in mobile communication, information processing, entertainment and defence have made them more so. We are now dependent on access to most of the periodic table and while the ores from which some of its members are drawn are plentiful, others are scarce, often localised in unsympathetic surroundings and controlled by regimes that may have other plans for them. And refining and synthesising materials is energy intensive – some 21% of all the energy we use is used to make materials.

The global population is increasing, and the affluence of this population is rising at the same time. With increased affluence comes increase in consumption, so, unless we can find ways to stop it, the consumption of materials and energy will rise considerably faster than the population itself. The vision expressed in the Brundtland Commission report – that of providing for the needs of the present without compromising the ability of future generations to meet their needs – is one that almost everyone would accept. But a consensus on how to achieve it is harder to achieve.

With this background we move, in the next chapter, to a closer look at what sustainable development involves.

1.8 EXERCISES

E1.1 What is the risk level assigned to Molybdenum (important for high-temperature alloys) in the British Geological Survey ranking of Material-criticality?

E1.2 The table lists the elements that appear in a present-day television set. Mark them up on the periodic table below, using color to distinguish between the critical (Table 1.1 of the text) and the noncritical elements. How many critical elements are in the set?

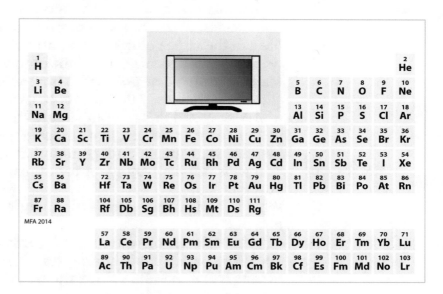

Elements in a Television

Aluminum	Iron	Rare earths
Antimony	Lead	Rhenium
Beryllium	Magnesium	Silicon
Chromium	Manganese	Strontium
Cobalt	Nickel	Tantalum
Copper	Niobium (Cb)	Tin
Gallium	Palladium	Titanium
Germanium	Phosphorus	Tungsten
Gold	Platinum	Vanadium
Indium	Potassium	Zinc

E1.3 What is the "resource curse"? Use the internet to assemble a short description.

E1.4 What does "resource imperialism" refer to? Use the internet to assemble a short description.

E1.5 What is the history and remit of the Intergovernmental Panel on Climate Change (IPCC)? Explore its web-site (www.IPCC.ch) to find out.

E1.6 All nations strive for economic growth, for the good reason that increased national wealth can enable higher quality of life. But the same policy can have undesired and often unanticipated consequences. List some of these.

E1.7 Many scientists and politicians now accept that the global climate is changing, that this is caused, in part, by industrial society and that nations should collaborate to counter it. But there are those that disagree. Use the internet to identify individuals and groups that argue that climate change is not happening, or is not man-made or is unimportant.

E1.8 What is a "circular materials economy"? Use the internet to find out.

E1.9 (a) The volume of water in the earth's oceans is about $V = V = 1.5 \times 10^9$ cubic kilometres. The surface area of the oceans is $A = A = 3.35 \times 10^8$ square kilometres. The temperature of this water varies from place to place and from surface to ocean floor. If, because of global warming, the temperature profile of the oceans shifts upwards by $\Delta T = \Delta T = 2^\circ$ C, by how much will the ocean surface rise because of thermal expansion the water? (The volumetric expansion coefficient of water is $\alpha_v = \alpha_v = 2.07 \times 10^{-4}/^\circ$C.) (b) If the floating ice cover in the Arctic also melts, what contribution will it make?

E1.10 The current (2014) growth rate of population in India is 1.6% per year. Its GDP is growing at 4.5% per year. Assuming that population P, affluence A (measured by GDP) and technology T are mutually independent, what change in resource intensity (meaning environmental impact per unit of GDP) must be achieved by technological development in India to ensure that environmental impact I gets no worse than it already is?

E1.11 Many Chinese cities have severe air pollution problems. The current (2014) growth rate of population in China is 0.5% per year. Its GDP is growing at 9.6% per year. Assuming that population P, affluence A (measured by GDP) and technology T are mutually independent, what change in resource intensity (meaning environmental impact per unit of GDP) must be achieved by technological development in China to ensure that environmental impact I gets no worse than it already is?

E1.12 Uganda is one of the world's fastest growing nations. Its population is growing at 4.5% per year and its GDP is growing at over 7% per year. Assuming that population P, affluence A (measured by GDP) and technology T are mutually independent, what rate of technological improvement will be required in Uganda to avoid an increasing impact on its natural resources?

Further Reading

Baker Susan: *Sustainable Development*, Routledge, 2013, ISBN: 978-1-134460-07-6 (A concise and readable explanation of sustainable development taking a global and national perspective.).

Carson, Rachel, (1962). "Silent Spring" first published by Houghton Mifflin, republished (2002) by Mariner Books. ISBN 0-618-24906-0. (The book, which argued that the uncontrolled use of pesticides was harming birds and animals, including humans, had an enormous impact.)

European Union Eurostat, (2013). http://epp.eurostat.ec.europa.eu/portal/page/portal/sdi/indicators (Tables of EU Indices of sustainability.)

IPCC: In Pachauri RK, Reisinger A, editors: *Climate Change 2007: Synthesis Report. Contribution of Working Groups I, II and III to the Fourth Assessment Report of the Intergovernmental Panel on Climate Change*, Geneva, Switzerland, 2007, IPCC. http://www.ipcc.ch (The report asserts that "most of the observed increase in global average temperatures since the mid-20th century is very likely due to the observed increase in anthropogenic greenhouse gas concentrations.").

IPCC (2014) IPCC, Geneva, Switzerland http://www.ipcc.ch (The Report, presented in Copenhagen, Denmark in 2014, will provide an update of knowledge on the scientific, technical and socio-economic aspects of climate change.)

Jaffe R, Price J: *Critical Elements for New Energy Technologies*, USA, 2010, American Physical Society Panel on Public Affairs (POPA) study, American Physical Soc.

Malthus TR: *An Essay on the Principle of Population*, Printed for London, 1798, Johnson, St. Paul's Church-yard. http://www.ac.wwu.edu/~stephan/malthus/malthus (The originator of the proposition that population growth must ultimately be limited by resource availability.).

Meadows DH, Meadows DL, Randers J, Behrens WW: *The limits to Growth*. New York, 1972, Universe Books (The "Club of Rome" report that triggered the first of a sequence of debates in the 20th century on the ultimate limits imposed by resource depletion.).

Meadows DH, Meadows DL, Randers J: *Beyond the Limits*, London, UK, 1992, Earthscan. ISSN:0896-0615. (The authors of "The Limits to Growth" use updated data and information to restate the case that continued population growth and consumption might outstrip the Earth's natural capacities.).

Mulder K: *Sustainable Development for Engineers: A Handbook and Resource Guide*, Sheffield UK, 2006, Greenleaf Publishing Ltd. ISBN: 13: 978-1-874719-19-9.

UNEP: *Report of the United Nations Conference on Human Environment*, Stockholm, 1972, The Earth Summit. http://www.unep.org/Documents.Multilingual/ Default.asp?DocumentID=97 (The Conference, known as the Earth Summit, put environmental issues on the international agenda for the first time and laid the groundwork for progress in the environment and development.).

UNEP: *The Montreal Protocol on Substances that Deplete the Ozone Layer*. http:// ozone.unep.org/new_site/en/montreal_protocol.php, 1989 (The most widely ratified treaty ever put before the United Nations, supported by 197 of the world's 210 nations.).

UNEP: *The Rio Declaration on Environment and Development*. http://www.unep. org/Documents.Multilingual/Default.asp?documentid=78&articleid=1163, 1992 (The Declaration lays out 27 "Principles" underpinning sustainable development, among them that "human beings are at the centre of concerns for sustainable development".).

UNFCC: *The Kyoto Protocol*. http://unfccc.int/kyoto_protocol/items/2830.php, 1997 (The Protocol set targets for the 44 industrialized countries that signed it, committing them to reduce greenhouse gas emissions over the five-year period 2008-2012. Few achieved their target.).

UNIDO: *The Stockholm Convention*. http://www.unido.org/index.php, 2001 (The Convention is a global treaty to protect human health and the environment from persistent organic pollutants (POPs)).

US Department of Energy: *Critical Materials Strategy for Clean Technology*. www. energy.gov/news/documents/criticalmaterialsstrategy.pdf, 2010 (A report on the role of rare earth elements and other materials in low-carbon energy technology).

US Department of Energy: *Critical Materials Strategy*, Office of Policy and International Affairs, 2010. materialstrategy@hq.doe.gov. www.energy. gov(A broader study than MRS 2010, above, but addressing many of the same issues of material critical to the energy, communication and defense industries, and the priorities for securing adequate supply.).

Circular 2112 USGS, Wagner LW: *Materials in the Economy –Material Flows, Scarcity and the Environment*, US Department of the Interior, 2002. www.usgs.gov (A readable and perceptive summary of the operation of the material supply-chain, the risks to which it is exposed, and the environmental consequences of material production.).

Visser W: *The Top 50 Sustainability Books* (The University of Cambridge Programme for Sustainability Leadership). Greenleaf publishing, 2009, ISBN: 978-1-906093-32-7(Four-page précis of key books on Sustainability, each listing Key ideas, a Synopsis and Author notes.).

WCED: *Annex to document A/42/427-Development and International Co-operation: Environment*, World Council on Economic Development, "Our Common Future" Oxford University Press and The United Nations, 1987 (The so-called "Brundtland Commission Report" clearly stating, for the first time, the need to balance economic growth with custody of the environment and social equity both today and for future generations.).

World Bank Development Indicators: http://data.worldbank.org/data-catalog/world-development-indicators, 2012 (The World Bank assembles annual figures for global economic development data at the national, regional and global levels.).

WSSD: *World Summit on Sustainable Development*. Johannesburg. http://www.who.int/trade/glossary/story097/en/index.html, 2002 (The summit identified five priorities for Sustainable Development: Water, Energy, Health, Agriculture and Biodiversity.).

CHAPTER 2

What is a "Sustainable Development"?

Materials and Sustainable Development. http://dx.doi.org/10.1016/B978-0-08-100176-9.00002-5
Copyright © 2016 Elsevier Ltd. All rights reserved.

Chapter Outline

2.1   Introduction and Synopsis   28
2.2   What Does "Sustainability" Mean?   28
2.3   Defining "Sustainable Development"   30

2.4   Articulations of Sustainable Development   33
2.5   Summary and Conclusions   36
2.6   Exercises   37
Further Reading   37

2.1   INTRODUCTION AND SYNOPSIS

"Sustainability" is an elusive term, one that (it sometimes seems) can mean whatever you want it to mean. A "sustainable development" is a little easier to pin down – it is any development that moves us from a less sustainable to a more sustainable state – but even this has a large number of dimensions. Three of the more important ones can be summarized as the "3Ps": prosperity, people and planet; we encountered them in Chapter 1 as economics, society and environment. In this chapter we re-interpret the 3Ps as three fundamental "capitals"; manufactured and financial capital, human and social capital, and natural capital. The *three capitals* form one of the building blocks of the method developed in the next two chapters. The other building blocks include the *objective* that motives the development, the *stakeholders* who are in some way involved or influenced by it, and the *facts* that characterize it. This chapter introduces these components. They come together in Chapter 3.

2.2   WHAT DOES "SUSTAINABILITY" MEAN?

What does "sustainability" mean? Conserving materials? Saving energy? Certainly. But there is more than that.

Take a look at the two images on the cover page of this chapter. The first, a train in Pakistan, depicts a mode of transport that is perhaps the most energy-efficient and material-efficient, per passenger mile, of any on the planet. And probably the cheapest. So energy, yes. Materials, yes. Economy, yes. But is it a *sustainable* solution for public transport? Probably not. There are issues of safety. It may not be legal. In more developed countries it might not be seen as socially acceptable.

The second image, a freeway in Australia, shows how those in a developed nation transport themselves. It is legal. It is relatively safe. It is socially acceptable. But is it a *sustainable* transport solution? Even auto makers begin to doubt it. It is extravagant with both energy and materials. It uses a great deal of space. It is expensive. It creates emissions that damage the environment and human health. It consumes oil, creating, for many nations, a dependence so powerful that it shapes foreign policies and political agendas and has, in the past, led to exploitation and human-rights violations. And, as the picture shows, it taxes human patience.

Now look back at the words in the last two paragraphs needed for this ultra-brief commentary on the sustainability of transport. Here are some of them:

Energy	Emissions	Exploitation
Materials	Environment	Efficiency
Space	Policy	Safety
Resources	Politics	Human patience
Economics	Legality	Social acceptance

So the first thing we have to recognize is that questions of "sustainability" do not have a simple "yes/no" answer. They have many facets – the efficient and conservative use of resources, environmental stewardship, rule of law, social equality, human rights, and more. Issues of sustainable development are intrinsically complex; their assessment requires acceptance of this complexity and ways of working with it. As we shall see, individual facets can be explored in a systematic way but the synthesis of these facets to give a final assessment requires reflection, judgment and debate. There is no simple, "right" answer to questions of sustainable development – instead, there is a thoughtful, well-researched response that recognizes the concerns of stakeholders, the conflicting priorities and the economic, legal and social constraints of the technology as well as its environmental legacy.

Before continuing it is helpful to introduce a distinction. "Sustainability" is an absolute term – something sustainable survives, something that is unsustainable does not. "Sustainable development" is a relative term: It is development that moves us from the present state towards a more nearly sustainable state. Thus the base-line is today's technology; the "development" refers to a change in that technology.

2.3   DEFINING "SUSTAINABLE DEVELOPMENT"

Here, repeated, is the much-quoted definition of sustainable development by the Brundtland Commission:

> Sustainable development is development that meets the needs of the present without compromising the ability of future generations to meet their own needs

The Brundtland Report of the World Council on Economic Development (WCED, 1987).

It sounds right. But how is it to be achieved? And where do materials fit in? The definition gives no concrete guidance.

So let us try another view of sustainability, one expressed in the language of accountancy: the triple bottom line or TBL (Figure 2.1). Almost any technical change has economic, environmental and social dimensions. The idea is that a corporation's ultimate success and health should be measured not just by the traditional financial bottom line, but also by its environmental and social and ethical performance. Instead of just reporting the standard bottom line of the income and outgoings ("prosperity") the balance sheet should also include the bottom lines of two further accounts: one tracking impact on the environmental ("planet") and one tracking the social ("people") balance. In this view, sustainable business practice requires that the bottom lines of all three columns show positive balances, represented by the "sustainable"

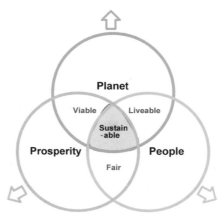

FIGURE 2.1
Triple bottom line, or "3P" thinking.

sweet-spot in the middle of Figure 2.1. Thus a corporation that is able to report an operating profit, improved training for its workforce and a reduction in energy use through more efficient practice can claim to have positive balances in all three bottom lines. The concept is a recent one – it originated just 20 years ago. Many businesses now implement TBL reporting; indeed the Dow Jones Sustainability Index of leading industries is based on it.

News-clip: "Triple bottom line – the 3 Ps: profit, people and planet. *The phrase "the triple bottom line" (TBL) was first coined in 1994 by John Elkington, the founder the British consultancy SustainAbility. Only a company that produces a TBL is taking account of the full cost involved in doing business."* The Economist, 17 November, 2009.

This, however, is sustainability analysis from the point of view of a corporation, or firm – a company-centric view. If we try to apply it to engineering design we run into difficulties. How does it help create sustainable development in the sense expressed by the Brundtland report? If the concern is that of global sustainability, with the welfare of both the people of today and of future generations, we need a broader perspective.

The three capitals. We make better progress if we separate the circles of Figure 2.1 and unpack their content, so to speak (Figure 2.2). Here a view of sustainability seen through the lens of

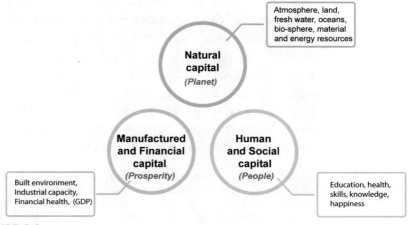

FIGURE 2.2
The three capitals

economics can help (Dasgupta, 2010). Global or national "wealth" can be seen as the sum of three components: the *net manufactured capital*, the *net human capital* and the *net natural capital*. They are defined like this:

- *Manufactured capital* – industrial capacity, institutions, roads, built environment and financial wealth.
- *Human capital* – health, education, skills, technical expertise, accumulated knowledge, happiness.
- *Natural capital* – clean atmosphere, fresh water, fertile land, productive oceans, accessible minerals and fossil energy.

All[1] can (with some difficulty) be quantified in a common measure, dollars, say. The sum of all three, the *net comprehensive capital*, is a measure of national or global wealth. "Strong" sustainability, in this picture, is development that delivers positive growth in all three capitals. "Weak sustainability" is development that delivers positive comprehensive capital, ensuring that the sum of the capitals passed on to future generations is positive, even if one of them is diminished. The risk in adopting the weak sustainability view is of an over-reliance on technology to compensate for a diminished ecosystem and depleted natural resources.

The main force, today, that drives change in the three capitals is the pressure for economic growth. An economy that grows is seen as "healthy"; one that does not grow is "stagnant" and one that is in recession is "sick". Positive economic growth is seen as so essential to the welfare of a nation that its influence on natural and human capitals is sometimes treated as secondary. Economic growth may contribute to human capital by enabling greater education and health care, for instance, or it may diminish it by encouraging unfair labor practices and social inequity. And unfettered economic growth must, in the long run, diminish natural capital by consuming irreplaceable resources.

---

[1]Both human and manufactured capitals have certain decay-time: people die, a new generation have to be educated; machines wear out and require maintenance or replacement. So maintaining even the present level of human and manufactured capital requires a flow of resources. Does the natural capital provide sufficient interest (via the ability to capture the sun's energy in plants and animals, and via the capacity of oceans and soil to absorb emissions) to provide the resources needed to maintain the other capitals?

2.4   ARTICULATIONS OF SUSTAINABLE DEVELOPMENT

Recognition of the importance of natural and social capital has stimulated activities to diminish the undesired impacts of economic growth on both – particularly to diminish resource consumption, emission-release and social inequity. These activities, of which there are many examples, are presented by their proponents as contributions to sustainable development. Each has a particular motivation. Here are some examples: to reduce the carbon emissions from cars, or to recover energy by incinerating waste, or to reclaim scarce elements from cast-off mobile phones. Following Mulder et al (2011) we will refer to them as "articulations" of sustainable development. The difficulty with almost all of them is that they conflict: An articulation that addresses one facet of the problem may aggravate another.

Figure 2.3 gives four examples. Each articulation is motivated by what we will call a *prime objective* (left-hand column). Achieving the prime objective is the desired outcome but the method chosen to do so can have unintended and undesirable consequences (right-hand column). Advocates of *bio-fuels* and *bio-polymers* do so because they diminish dependence on fossil hydrocarbons, but the land and water needed to grow the necessary feedstock is no longer available for cultivating food. *Carbon taxes* are designed to stimulate a low-carbon economy, but they increase the price of

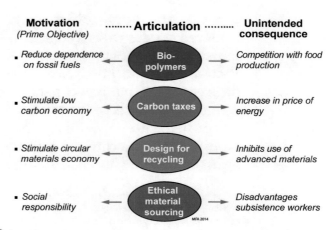

FIGURE 2.3

Examples of articulations of sustainable development. Each has a motivating objective. Many also have unintended consequences.

energy and hence of materials and products. *Design for recycling* is intended to meet the demand for materials with less drain on natural resources, but it constrains the use of light-weight composites because most cannot be recycled. The motivation for *ethical sourcing of raw materials* (sourcing them from nations with acceptable records of human rights) is that of social responsibility but a side-effect is to increase the prices paid for resources and perhaps to deprive subsistence-level workers of work. The many different articulations of sustainable development of technology aim to support one or another of the three capitals of Figure 2.2. But while many address one facet of the problem, few support all three.

We have examined some 80 articulations of sustainable development, drawn from journals[2] that specialize on the subject, identifying the prime objective of each. Figure 2.4 is a map showing some of them. It is a schematic rather than an accurate representation. The words in the bubbles are generic – each is a headline, so to speak, for a family of articulations. The bubbles are color-coded according to their nature. They are not independent and may not be mutually supportive.

What generalizations – meta-messages – can be distilled from collections like that of Figure 2.4?

- Each of these articulations has a motivating target that we called its *prime objective*. Each, generally, has a *size-scale*, a *time-scale* and a *cost-scale* that are envisaged for its implementation. The first step in assessing an articulation is to identify these. If the prime objective is unrealistic or cannot be achieved on the envisaged scales, the articulation is not sustainable in its present form but will require modification.
- Each articulation involves a set of *stakeholders*. Stakeholders have concerns that reflect perceived environmental, social or economic impacts of the articulation. If compromise cannot be reached with powerful and concerned stakeholders, it will be difficult to execute the articulation in an acceptable way, again compromising sustainability.

[2]World Journal of Science, Technology and Sustainable Development, ISSN: 2042-5945. World Journal of Science, Technology and Sustainable Development, ISSN: 2042-5945. Journal of Technology Management & Innovation, ISSN 0718-2724. Journal of Clean Technology and Environmental Sciences, ISSN: 1052-1062. Clean Technologies and Environmental Policy, ISSN: 1618-9558; Proceedings of the National Academy of Science, www.pnas.org/

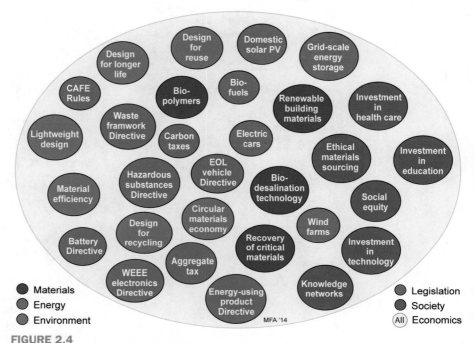

FIGURE 2.4
A few of the many articulations of sustainable development with the sector on which they act most strongly. Most articulations have some impact on all six sectors.

We need to examine these obstacles more carefully. Sustainability, however, means more than achievability. We need to deconstruct the articulations of Figure 2.4 further. As suggested in the figure, the central issues can be grouped under six broad headings summarized below and in Figure 2.5:

- *Materials*: whole-life material efficiency, supply-chain security.
- *Energy*: whole-life energy efficiency, security of energy supply.
- *Environment*: resource consumption, emissions to air, water and land.
- *Regulation*: awareness of, and compliance with, national and international agreements, legislation, directives, restrictions and agreements.
- *Society*: health, education, shelter, employment, equity, knowledge, intellectual property, happiness.
- *Economics*: the cost of implementing the articulation and the benefits that it might provide.

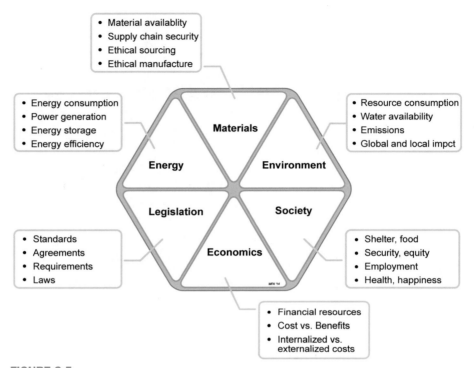

- Material availablity
- Supply chain security
- Ethical sourcing
- Ethical manufacture

- Energy consumption
- Power generation
- Energy storage
- Energy efficiency

- Resource consumption
- Water availability
- Emissions
- Global and local impct

Materials

Energy   **Environment**

Legislation   **Society**

- Standards
- Agreements
- Requirements
- Laws

Economics

- Shelter, food
- Security, equity
- Employment
- Health, happiness

- Financial resources
- Cost vs. Benefits
- Internalized vs. externalized costs

FIGURE 2.5
The six major sectors that are involved in most articulations of sustainable development.

Each heads a check-list for what might be called "sustainability analysis" of a design, scheme, project or product. While one may be dominant in a particular articulation, the others almost always play a role too.

2.5   SUMMARY AND CONCLUSIONS

So what is the picture now? At one extreme we have the three capitals – the 3P's if you prefer. They and their sum, the comprehensive capital, form the high-level landscape of a sustainable development. At the other extreme we have individual articulations of sustainable technical development: projects with a prime objective and a scope and timing. In between lie the stakeholders with their concerns and the hard facts that characterize the articulation itself. Linking them up is the challenge. The next chapter outlines how this is done. The one after that fills in the details.

2.6 EXERCISES

E2.1 "Growth" is seen as desirable in a market economy because it contributes to manufactured and financial capital. Can you think of things that contribute to one or other of the capitals by *not* growing, or, better, by shrinking?

E2.2 The global financial crisis that started with the collapse of Lehman in 2008 had impacts on all three capitals. Can you suggest at least one impact for each?

E2.3 The Battery Directive – an EU Regulation – requires that all batteries must be collected and recycled. The primary objective here is to prevent the contamination of landfill sites by heavy metals (lead, cadmium, mercury). What consequences for each of the three capitals do you think follow from this legislation?

E2.4 The US Dodd–Frank Act of 2010 forbids trade in material resources and commercial products by US companies with nations suspected of human rights abuses and of supporting armed conflict. What consequences for each of the three capitals do you think follow from this legislation?

E2.5 Fracking (hydraulic fracturing) of deep shale beds to release gas is seen by some as an articulation of a sustainable development. Debate this from two standpoints: that of the techno-optimist and that of the eco-pessimist, the first advocating "strong" sustainability, the other advocating "weak" sustainability.

Further Reading

Ashby MF: *Materials and the Environment – Eco-informed Material Choice*, Oxford, UK, 2012, Butterworth Heinemann, ISBN: 978-0-12-385971-6 (An introduction to the environmental aspects of the use of materials.).

Ayres R: *Viewpoint: Weak versus Strong Sustainability*. http://kisi.deu.edu.tr/sedef. akgungor/ayres.pdf, 1998.

Brundlandt D: *Report of the World Commission on the Environment and Development*, Oxford, UK, 1987, Oxford University Press, ISBN: 0-19-282080-X (A much quoted report that introduced the challenges and potential difficulties of ensuring a sustainable future.).

Dasgupta P: Natures role in sustaining economic development, *Phil. Trans. Roy. Soc. B* 365:5–11, 2010 (A concise exposition of the ideas of economic, human and natural capital.).

Dow Jones Sustainability Index, 2012. http://www.sustainability-index.com/. (One of several indices of sustainability, reviewed in Chapter 4.)

Elkington J: *Cannibals with Forks: The Triple Bottom Line of 21st Century Business*, Stony Creek, CT, 1998, New Society Publishers (Oddly named but interesting introduction to TBL idea.).

Mulder K, editor: *Sustainable Development for Engineers: A Handbook and Resource Guide*, Sheffield UK, 2006, Greenleaf Publishing Ltd. ISBN:13: 978-1-874719-19-9. (A set of essays that aims to give engineering students insight into the challenge of sustainable development and the potential contributions of engineering.).

Mulder K, Ferrer D, Van Lente H: *What is Sustainable Technology*, Sheffield, UK, 2011, Greenleaf Publishing, ISBN: 978-1-906093-50-1 (A set of invited essays on aspects of sustainability, with a perceptive introduction and summing up.).

Norman W, MacDonald C: Getting to the bottom of the "triple bottom line, *Bus. Ethics Q.* 14(2):243–262, 2004. http://isites.harvard.edu/fs/docs/icb.topic549945.files/Canadian%20Paper.pdf.

Pava M: A Response to "Getting to the Bottom of 'Triple Bottom Line'", *Bus. Ethics Q.* 17(1):105–110, 2007.

Porritt J: *The Five Capitals.* http://www.forumforthefuture.org/project/five-capitals/overview, 2012.

UN Global Compact, 2012. www.unglobalcompact.org/.

WCED: *Report of the World Commission on the Environment and Development*, Oxford, UK, 1987, Oxford University Press (The Brundtland Committee report.).

Weaver P, Jansen L, van Grootveld G, van Spiegel E, Vergragt P: *SustainableTechnology Development*, Greenleaf Publishing Ltd, 2000, ISBN: 187-471-9098. Accessible in Google books here (See particularly Chapters 2 and 3).

Wiek A, Lauren Withycombe L, Redman CL: *Key Competencies in Sustainability: A Reference Framework for Academic Program Development*, 2011.

CHAPTER 3

Assessing Sustainable Developments: The Steps

Chapter Outline

3.1 Introduction and Synopsis 39
3.2 Dealing with Complex Systems 40
3.3 A Layered Approach to Assessing a Sustainable Development 42
3.4 Assembling the Layers 50
3.5 Summary and Conclusions 51
3.6 Exercises 51
Further Reading 54

3.1 INTRODUCTION AND SYNOPSIS

The last chapter introduced some of the facets of an articulation of sustainable development. An articualtion has a prime objective – something it is aiming to do or change. It has stakeholders, interested parties, some enthusiastic, some reluctant. And it has a factual background that can be researched. But how do these, collectively, bear on the three capitals and, ultimately, on establishing the articulation as a sustainable development? To answer that we need a method.

Materials and Sustainable Development. http://dx.doi.org/10.1016/B978-0-08-100176-9.00003-7
Copyright © 2016 Elsevier Ltd. All rights reserved.

This chapter explains the method. It reviews the components and assembles them into a framework. It ends with a schematic of this framework. It includes a running example that shows how the steps work but is not otherwise to be taken seriously. The chapter is a sort of executive summary, providing an overview of the approach. Like most executive summaries, it paints the big picture but leaves out much detail. Chapter 4 provides the detail.

We are dealing here with complex systems. "Complex" because many disparate entities are involved; "System" because the entities interact. So first, a brief digression on thinking about complex systems.

3.2 DEALING WITH COMPLEX SYSTEMS

It is natural to feel uncomfortable when confronted with problems that are multi-dimensional, interactive and poorly defined. The answer is to have a framework for critical thinking that recognizes the complexity and the interdependence and allows you to work with them.

One approach is to split the problem into layers (Figure 3.1). The bottom layer, the starting point, is a statement of the problem. Problems have a context: the circumstances that surround them. Why and how has the problem arisen? What outcomes would be desirable? These form the second layer. Now factual information about the problem and its context can be researched in a systematic, value-independent way. Technical, economic and legal aspects, for instance, lend themselves to objective research. No implications are sought here; the facts are simply assembled while suspending all judgment. The facts are stored on the third layer.

Complex problems would not be complex if systematic research alone could solve them. The complexity usually lies in the interdependence of what are sometimes called "incommensurate quantities" – things that are measured in different units or are not "measured" at all: personal judgments, culture or belief-dependent values. To move from the third layer to the fourth, that of value-based assessment, may require recognition of diverse views, only resolvable by discussion, debate and compromise to reach a mutually acceptable position.

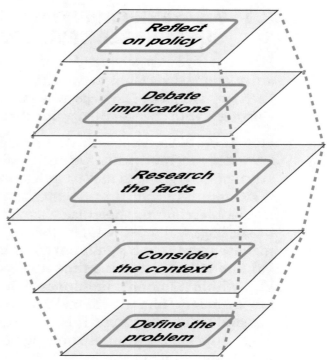

FIGURE 3.1
Layering as a way of thinking about complex problems.

The final step is one of reflection. What conclusions for strategy or action can be drawn from the debate? It is possible that any solution to the problem will leave some of the parties involved dissatisfied. Are there ways to involve them in ways that reduce the dissatisfaction?

This layer-based approach clearly separates the objective, fact-based aspects of the problem from the more difficult value-based aspects. It allows thinking about interaction *within* each layer, followed by interaction *between* layers. Broadly speaking the lower layers inform the ones above, so that the approach has a sequence and a direction (from bottom to top). That is not to say that it is linear – thinking about any one layer may require further clarification of the layers below. But it does give a framework.

Let's now see how it might play out in analyzing an articulation of sustainable development.

3.3 A LAYERED APPROACH TO ASSESSING A SUSTAINABLE DEVELOPMENT

Here is a five-step strategy for assessing a design or project (an articulation) that claims to contribute to sustainable development. Each step is a layer. The steps are summarized here and detailed in Chapter 4 with check-lists to guide their implementation.

Step 1: Problem definition. Any articulation of sustainability has an underlying motive that we earlier called its *prime objective*. If the articulation is going to make a difference it must act on a scale that is significant in comparison with that of the problem itself. Thus legislation requiring supermarkets to provide only bio-degradable plastic bags will make a difference only if plastic bags from supermarkets constitute a significant fraction of *all* plastic bags. Similarly, an articulation has a time scale. Insisting on bio-degradable bags within 12 months presupposes that the supply chain for the bio-degradable film used to make them can cope with the resulting demand within that time. It is not possible to judge the viability of the articulation without knowing how large it will be and how soon it should happen (Figure 3.2).

Example. The Beneficial Brewing Corporation markets beer in 16 ounce (473 ml) aluminum cans. Their sales average 500 million cans per year, roughly 1% of the US beer market. At the annual general meeting (AGM) a group of eco-minded shareholders propose that the company should use steel cans instead of aluminum because steel has a lower embodied energy and carbon footprint than aluminum (Figure 3.3). The CEO of Beneficial Brewing is thereby presented with an articulation for a sustainable development. Here it is

- *Prime objective:* reduce energy demand and carbon emission by replacing aluminum by steel cans.
- *Size scale:* 500 million cans per year.
- *Time scale:* less clear, but the shareholders will expect some sort of response by the time of the next AGM in 1-year time.

We will need a size and time scale for the case studies later on. As in this example, the original statement of an articulation is often vague about these, yet they are always there. If they are not explicit, we will infer sensible default values from the context.

> **Prime Objective**
> - Motivation
> - Scale
> - Timing

FIGURE 3.2
The articulation.

FIGURE 3.3
An aluminum beer can.

The first step, then, is to *clarify the prime objective and its scale, physical and temporal.*

Step 2: Identify stakeholders and their concerns. Stakeholders are individuals, groups or organizations that are in any way affected by the articulation. Some, like the originators of the articulation in question, wish to see it succeed. Others may have reservations or voice outright opposition. It is important to identify the stakeholders and their concerns. If the concerns are not addressed the articulation will face obstacles and may fail to gain acceptance. If this happens the articulation is not sustainable.

How are stakeholders identified? A simple check-list, introduced in the next chapter, acts as a prompt. The National Press can provide background: controversial articulations (building land-based wind farms, for instance, or fracking for shale gas) cause stakeholders to express their concerns through Editorials, News and Business reports, Letters to the Editor and commentaries in the press and on radio and television – numerous examples appear in the case studies later in this book. Ultimately, however, stakeholder concerns are best identified by face-to-face meetings, phone interviews or questionnaires.

Stakeholders differ in their level of interest and the influence or power that they can exert. Figure 3.4 is a diagram with stakeholder interest and influence as axes. The likely behavior of a particular stakeholder depends, to some extent, on the position they occupy on this diagram – we go into details in Chapter 4. Once positioned, it is possible to reflect on the mutual influence or dependence of the stakeholders, shown here by arrows.

Example, continued

The CEO of Beneficial Brewing asks: who is interested or affected if we change from aluminum to steel cans? The shareholders have urged the change and are in a position to exert pressure on the company to adopt it: they are stakeholders with both interest and influence. The makers of aluminum cans may not wish to lose trade, but the makers of steel cans may be happy to get it – both are interested parties. Surveys suggest that most beer drinkers do not know or care what the cans are made from – they are stakeholders with little interest or influence so long as they get their beer. Law makers could, if so motivated, pass legislation mandating the use of steel cans but there is little reason to think that they would; they have influence but no interest. The important stakeholders are those above the diagonal (dotted) line. This is useful information, focusing the attention of the CEO on the key players and their concerns. Their views must be recognized in seeking the best path forward.

FIGURE 3.4
The stakeholder diagram for the CEO of the beneficial brewer, with paths of influence.

The second step, then, is to *identify the stakeholders and their concerns –* they set the context in which the assessment is carried out.

Step 3: Fact finding. To get further we need facts and facts need research. What sort of facts?

- Facts that establish what the articulation is and what materials and energy are required to make it happen. What environmental impact will it have? Is it legal? Are there regulations with which is must comply? Is it fair and equitable? What will it cost?
- Facts that relate to the stakeholder concerns. Are the concerns justified? What information is needed to confirm or refute them?
- Facts relating to essential infrastructure. What products or services will have to be in place to support the articulation if it goes ahead?

Each of these questions can be researched in an objective way using generally available sources: books, databases, interviews and the Internet, guided by check-lists.

What facts would be helpful to the CEO of Beneficial Brewing? The survey of sustainable development projects, described in Chapter 2, highlighted six groups of relevant information. They are shown in the six segments of Figure 3.5.

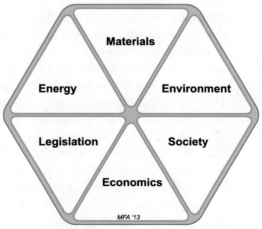

FIGURE 3.5
Fact-finding – an aid-memoire.

■ **Materials.** Is the supply-chain secure? Is a supplier of steel cans available? Have they the spare capacity to provide 500 million cans per year?

■ **Energy.** The shareholders believe that steel cans require less energy than aluminum cans. What are the values? If the change was made, how much energy would the company save in a year? What is this as a fraction of the total energy used by the company? Is it significant?

■ **Environment.** What are the relative environmental impacts of the two sorts of can? Does one have a lower carbon footprint than the other? Is one recycled more effectively than the other?

■ **Regulation.** What regulations bear on the use and recycling of cans? Is it the same for steel and aluminum? Are there any other legal or regulatory constraints?

■ **Society.** Are steel cans as acceptable as aluminum to drinkers of beneficial beer? Would the lower embodied energy of the cans be seen as a demonstration of environmental responsibility?

■ **Economics.** Do steel cans cost less than aluminum cans? What is the cost of re-equipping to cope with the change from steel and of aluminum? What are the benefits? Do they justify the cost?

Example, continued

Material, Environment and Energy. It is true, as the shareholders of the Benevolent Brewing Co claim, that steel has a much lower embodied energy than aluminum for virgin material – it is about 25 MJ/kg for steel, 200 MJ/kg for aluminum, a factor of 8 larger. Cans are not made from virgin stock but from stock with a considerable recycle content. The embodied energy of typical grades of can stock are about 18 MJ/kg for steel and 110 MJ/kg for aluminum, a factor of 6. Does this mean that the embodied energy of the two sorts of can differ by the same factor? The answer is no – a 5000-series aluminum 440 ml can weighs 13 grams; the equivalent steel can weighs 44 grams (Figure 3.6), so the embodied energies per can differ by much less – that of the aluminum can is just 1.7 times more than the steel one.

The forming energies to make the cans also differ. To make a valid comparison the CEO of Beneficial Brewing needs a life cycle assessment (LCA) for the production of each type of can. A detailed LCA from 2002[1] reaches the conclusion that the differences both in energy and in carbon emissions for the two types of can are so small that, given the inherent uncertainty in all embodied energy data, the energies and carbon emissions of the two are not significantly different.

Legislation and Regulation. Much regulation, easily found via a Web-search, now applies to packaging such as cans. The UK packaging (Essential Requirements) Regulations of 2003 is typical. It applies to any company that makes, fills, sells or handles packaging. It aims to minimise waste and ensure that packaging can be reused, recovered or recycled. To comply, a producer must join a registered compliance scheme. The legislation applies equally to aluminium and steel cans.

Economics. Can-grade steel costs about $0.4/kg; so the material cost for 500 million steel cans is about $8.8 million. Can-grade aluminum costs about $1.7/kg, making the material cost for 500 million cans $11.0 million. There could, therefore, be a possible saving of $2.2 million in changing to steel.

Society. Are steel cans as acceptable to the beer drinking public as aluminum cans? Surveys suggest that most do not care, and the fact that the two competing brands pictured in Figure 3.6 use different can materials reinforces this perception.

[1]http://www.apeal.org/uploads/Library/LCA%20study.pdf

FIGURE 3.6
440 ml aluminum (left) and steel (right) cans.

The third step, then, is one of fact-finding, temporarily putting all judgment on hold.

Step 4: Synthesis. The fourth step, *synthesis*, is one of drawing together the facts from Step 3 to form a balanced judgment about the impacts on the three capitals (Figure 3.7). Each capital has a base-value – its value today. The articulation, if implemented, will change these values. It will be the job of the CEO at the next AGM to guide the debate towards a rational assessment of these changes.

It is here that values, culture, beliefs and ethics enter more strongly. As mentioned in Chapter 1, an environmentalist might argue that the impact on natural capital ranked most highly: after all, the natural environment is the support system of all life. Humanists might see understanding, reason, humanity and happiness as the central pillars of a civilized society and feel that any impact on human capital was unacceptable. To an economist, economic stability and growth of manufactured capital could seem to be the first priority, arguing that these provide the resources needed to protect the environment, enable innovation and support a vibrant society. Each of these groups recognizes the cases made by the others; indeed they have many concerns in common. But their final judgment will be influenced by their underlying beliefs and values, cultural, religious and political. It is no surprise that one set of facts can be interpreted in more than one way.

All this must be accepted. The important things to retain

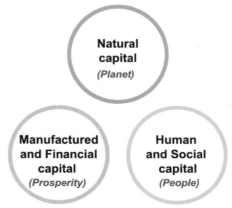

FIGURE 3.7
Synthesis – debating the impact of the facts on the three capitals.

- Respect for the facts
- Respect for alternative interpretations of the facts
- Respect for the value of compromise reached by reasoned debate.

Example: continued

The CEO can now present the facts to the Brewery Board and initiate a discussion of their impact on the three capitals.

Natural capital. Contrary to the intuition of the shareholders, the facts suggest that the differences in embodied energy and carbon footprint of steel and aluminum cans are too small to be significant. This is because of the high recycle content of can-stock, because aluminum cans are much lighter than those made of steel and because (according to the LCA) the deep-drawing of aluminum to make cans is less energy intensive than the equivalent process for steel. The supply chains for both metals are robust with no global or national shortages (indeed at the time of writing there is over-capacity). Beneficial Brewing requires only 1% of the can market and cans account for about 10% of the global aluminum consumption so the impact of material choice by Beneficial is very small.

Human capital. A can is … well … just a can. The material of which it is made carries no emotional, cultural or (since it is decorated) intrinsic aesthetic baggage that needs unpacking. No significant impact here.

Manufactured and financial capital. If the prices of steel and aluminum are directly reflected in can prices, a switch to steel could provide an annual saving of about $2 million. At a (guessed) shipping price of 50¢ per filled can, Beneficial's revenue stream from beer is of order $250 million, so this saving is about 0.8% of turnover. But against this must be set the cost of re-equipping the brewery's production line to deal with steel cans and the possible disruption of production while this happens. The CEO and the Board take the view that the risks exceed the benefits.

Step four, Synthesis, is the most difficult one. The next chapter suggests a matrix that can help with it.

Step 5: Reflection. The fifth and last step is that of *reflection on alternatives*. Is the prime objective achieved? Is it achieved on a scale that makes a significant difference? Do the benefits to the three capitals outweigh the negative impacts? Can the analysis suggest a new, more productive, way of achieving the prime objective?

Example, continued

Is the shareholders' "articulation" a sustainable development or not? Taken together, the impacts on the three capitals suggest that it is not. But the shareholders are stakeholders with both interest and influence. Their views must be respected.

Reflection. This is the moment to return to the Beneficial Brewery, pour a glass of beer, and ponder on alternatives – preferably those that do not require re-equipping the production line. The prime objective was to reduce depletion of natural capital associated with beer cans. Could aluminum cans be made thinner and thus less energy intensive? Aluminum can makers have already thought of that. Increase the recovery of aluminum cans for recycling by charging a deposit? That will work only if it is mandated nation- or state-wide, something the brewery cannot do by itself. But the brewery could lobby for such legislation, thereby demonstrating to shareholders its commitment to the environment without the disruption of changing material.

3.4   ASSEMBLING THE LAYERS

The layers are stacked in Figure 3.8 in the ascending sequence:

- Articulation statement
- Stakeholder and their concerns
- Fact-finding
- Synthesis: interpretation of the facts
- Reflection

The lower layers inform the ones above. As explained earlier, the layer-based approach clearly separates the objective, fact-based aspects of the problem that can be explored in a systematic, scientific way from the more difficult value-based aspects. It encourages thinking about interaction within each layer, and gives a logical path to explore the interaction between layers. Figure 3.8 is both a visualization and a summary of the method, one that is followed more or less exactly in the six case studies of Chapters 8–13. Before moving onto those, we examine the steps shown in Figure 3.8 in more detail (Chapter 4).

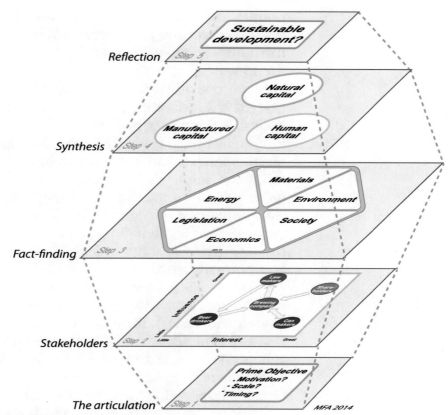

FIGURE 3.8
The layered approach to analyzing an articulation of sustainable development.

3.5 SUMMARY AND CONCLUSIONS

There is no completely "right" answer to questions of sustainability development; instead there is a thoughtful, well-researched response that recognizes the many conflicting facets and seeks the most productive compromise. The layer-based approach described here and developed further in Chapter 4 provides a framework for doing this. The method is designed to help teachers introduce students to sustainability analysis in a simple, progressive way. The case studies of Chapters 7–13 illustrate it in operation.

3.6 EXERCISES

E3.1 Energy supply in Germany is principally derived from fossil fuels. Electricity consumption in Germany is at present about 600TWh per year. Germany's energy policies and the politics of "Energiewende" involve the phasing out of nuclear power and progressive reduction of dependence on fossil fuels by replacing them by renewables. The plan calls for 35% of all electricity generation to come from renewable sources, principally wind, by 2020. That is an articulation of a sustainable development. What are the prime objective, the timing and the scale?

E3.2   Biopolymers are promoted as sustainable substitutes for plastics derived from oil. The bio-based Society[2] makes the following claim: "Biobased polymers will increase in capacity to 12 million tons/year by 2020. That will equal 3% of total polymer production." That is an articulation of a sustainable development. What, concisely, are the prime objective, the timing and the scale?

E3.3   You normally drive to work – a distance of 5 km (3 miles) and do so 250 days per year. You decide to be more" sustainable" by cycling instead (an articulation of sustainable development). Express this as a prime objective and a size and time-scale. The distance driven by an average person per year in Europe is about 20,000 km. Use this to assess the relative scale of the articulation.

E3.4   Cars are responsible for about 15% of carbon emissions in a developed country. The agenda of one political party in the UK is to introduce legislation requiring that only electric cars will be allowed on the British roads from 2030. Formulate this as a prime objective, scale and timing. (Approximately 2.5 million cars are sold in Britain each year.)

E3.5   Coffins, traditionally, are made of hardwood with bronze fittings and they are expensive. It is suggested that a more sustainable solution might be a coffin made from recycled cardboard. The Green Lobby thinks it a good idea but has little influence; they can, however, lobby the coffin makers and coffin-users. Opposition may, however, be expected from the Worshipful Guild of Coffin-Makers (a Union) who see cardboard coffins as a threat to the profitability of their trade. They can put pressure on coffin-users to bury their loved ones in something more dignified. Funeral home directors, too, are

---

[2]http://www.biobased-society.eu/2013/03/biobased-polymers-will-make-a-breakthrough-within-ten-years/.

involved – the margin on providing a cardboard coffin is likely to be smaller than that of a wooden one. These diverse views need to be recognized and a strategy sought to bring all stakeholders on board. Make a stakeholder interest/influence diagram and position these stakeholders on it. Then add any others that strike you as relevant.

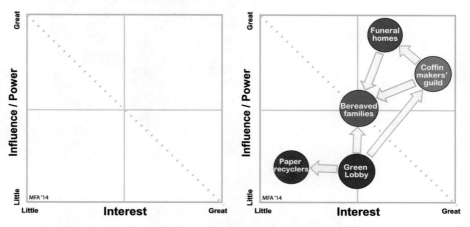

E3.6. Solar chargers for mobile phones.

"If chargers for devices such as mobile phones and MP3 players were unplugged when not in use, the UK could save enough electricity to power 115,000 homes, about 2 GJ per year" – the BBC's advice on saving energy.

But why not save even more energy by not plugging them in at all? Use a solar charger instead. The one in the picture has an internal 800 mAh lithium-ion battery with a life of 2 years. It weighs 49 grams. In using it you are "helping to do your bit to save the planet" – a product review.

Translate this into an articulation of sustainable development. In purely energy terms, is this articulation, in reality, a sustainable one? If there is a net energy saving, is it significant? This is an exercise in approximate (order of magnitude) analysis.

It helps to know that a typical mobile phone uses 4 kWh per year, and that a phone charger, plugged in but not in use, draws about 0.5 W so if it is permanently plugged in while not charging, it uses about another 4 kWh per year. The embodied energy of small battery-powered electronics average, very approximately 0.4 MJ of primary energy per gram.

Recycling of hand-held electronics is not yet widely practiced. The average UK resident uses about 6000 kW.h of electrical energy per year.

Further Reading

Mulder K, Ferrer D, Van Lente H: *What is Sustainable Technology*, Sheffield, UK, 2011, Greenleaf Publishing, ISBN: 978-1-906093-50-1 (A set of invited essays on aspects of sustainability, with a perceptive introduction and summing up).

CHAPTER 4

Tools, Prompts and Check-Lists

Natural capital

Manufactured capital

Human capital

Chapter Outline

4.1 Introduction and Synopsis 56

4.2 Step 1: Clarifying the Prime Objective 56

4.3 Step 2: Stakeholder Analysis 57

4.4 Step 3: Fact-Finding 60

4.5 Step 4: Informed Synthesis 66

4.6 Step 5: Reflection on Alternatives 71

4.7 Summary and Conclusions 71

Further Reading 72

4.8 Appendix: Creativity Aids – A Brief Survey 72

Further Reading: Creativity Aids and Reasoning 80

4.9 Exercises 82

Materials and Sustainable Development. http://dx.doi.org/10.1016/B978-0-08-100176-9.00004-9
Copyright © 2016 Elsevier Ltd. All rights reserved.

4.1 INTRODUCTION AND SYNOPSIS

Chapter 3 outlined the five-step approach to assessing a proposed articulation of a sustainable technological development. This chapter adds the details, explaining how to tackle the steps. We start with the *primary objective and its scale*. This is followed by *stakeholder analysis*. With these two steps in place, it is possible to start the search for facts. *Fact-finding methods* are described. We then examine the *synthesis* of the impact of the facts on the three capitals. That puts us in a position to *reflect* on a balanced judgment of the viability (or non-viability) of the project and of alternatives. The first three steps follow a systematic procedure. The last two – synthesis and reflection – involve judgment and the ability to think on a broad scale. Guidance is provided by check lists, prompts and cause-effect matrices.

This chapter is intended in part for reference. Fully worked case studies using the five-step method are introduced in Chapter 7 and developed in Chapters 8 to 13.

4.2 STEP 1: CLARIFYING THE PRIME OBJECTIVE

Any articulation of sustainable technology has a motive and a scale. The size-scale, often, is large – it has to be if the articulation is going to make a significant difference. The time allocated for implementation is equally important. Typically it is set by strategic government planning or by economic targets. If the articulation is to be implemented quickly (in a year, say) there may be insufficient time to set up the infrastructure needed to support it. If instead the time scale is very long (say 50 years) stakeholders may lose commitment.

A large scale and a fixed timing may mean that the articulation, if implemented, will demand resources on a correspondingly large scale and have significant economic, social and political implications. Thus building wind turbines to provide grid-scale electrical power will contribute in a significant way only if it provides a significant part – 5%, say – of national or global power, but to do this will require large quantities of materials and space. The high density of turbines needed to meet this target may be socially unacceptable, and the investment will be large. Anticipating the consequences of the scale and timing of an articulation on material, economic and social resources is a component in analyzing its viability as a sustainable development.

Table 4.1	Examples of Articulations, with Prime Objective, Size and Time Scales		
Articulation	**Prime Objective**	**Scale**	**Timing**
Biofuel additions to conventional auto fuels	Reduce dependence on fossil fuels	10% of all auto fuel by volume	By 2020
Mandatory charge for plastic bags	Reduce plastic pollution by discarded bags	Applies to all supermarket purchases	By 2015
Battery directive: mandatory recycling of all batteries	Reduce toxic metal contamination in landfill	All batteries, Europe-wide	By 2020
Waste electrical and electronic equipment directive	Reduce waste from obsolete electrical equipment	All electronics, Europe-wide	Now
End of life vehicles directive	Increase recycling of old vehicles	85% of every vehicle to be recycled, Europe wide	By 2015
Feed-in tariff – a subsidy for domestic solar power	Increase renewable electrical power generation	Available for all domestic solar installations	For 25 years from date of installation
Product take-back legislation for vehicle tires	Shift waste-management costs from state to tire producers	All tires, Europe wide	Now
Tidal power generation in the Swansea Bay, South Wales	Regional low carbon electric power generation	10-km barrier harvesting 10^6 MWh per year	By 2023

Table 4.1 lists the prime objectives of some national-scale articulations that are in progress or are proposed in Europe. If the prime objective is not achievable on the scale and within the time envisaged, the success of the articulation is compromised and it will need reconsideration. Thus the first step is to express the articulation as an objective with a defined size scale and allocated time.

4.3   STEP 2: STAKEHOLDER ANALYSIS

If you have interest in a particular project, then you hold a stake: you are a stakeholder. If you wish to participate in and change that project, interest is not enough; you also need influence. Stakeholder analysis means identifying the interested parties – the stakeholders – and

Table 4.2 Possible Stakeholders		
Local or National Government	Suppliers	Customers, Existing and Potential
Owners	The public or local community	Lobbyists and interest-groups
Employees	Trade unions	Investors, shareholders
Health and planning authorities	The press, radio and television	Managers, colleagues or team
Alliance partners	Business partners	The scientific community

their concerns, their influence and the ways they interact. The three key stakeholder questions are:

- Who are they?
- What do they want?
- How will they try to get it?

Failure to identify, respect and involve the stakeholders in a project is likely to generate opposition that may obstruct or defeat the purpose of the project itself.

Stakeholders may be both organizations and people (Table 4.2). How are they identified? Channels exist for stakeholders to identify themselves and publicize their concerns; they include editorials, interviews and letters in the national press, radio and television, focus-group reports, shareholder meetings, manifestos, demonstrations, interventions and in more extreme cases, vetoes, boycotts and strikes. The Internet allows many of these to be identified and explored. Beyond this, structured and semi-structured interviews and questionnaires provide some of the necessary information. Snowball sampling (asking known stakeholders for pointers to further stakeholders) and social network analysis expand the area of exploration further.

The stakeholder interest and influence matrix that we met in Chapter 3 (Figure 3.4) helps clarify the position and relationships of stakeholders. Figure 4.1 (left) elaborates. Stakeholders with little interest and little influence (*bystanders* in the figure) are unlikely to demand attention, though it is prudent to keep them informed. Stakeholders with great interest and great influence (*key players*) require full involvement with decision-making; achieving their active support may require negotiation and compromise. Stakeholders with great influence but little interest (*context-setters*) need

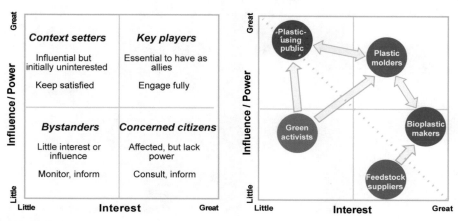

FIGURE 4.1

(Left) The components of the stakeholder interest and influence matrix. (Right) A stakeholder diagram for promotion of bio-plastics. The arrows indicate influence.

attention; they could be drawn in by other, less influential, stakeholders. Finally there are stakeholders with great interest but, individually, little influence (*concerned citizens*); if marginalized, their frustration may drive them to form alliances with each other and with more influential groups.

Stakeholders can be mapped onto the matrix, locating their interest and influence. Figure 4.1 (right) is one possible scenario. The articulation is that of promoting bio-plastics to reduce dependence on oil-based plastics. The suppliers of feedstock (maize, corn) have relatively little influence – they depend on the bioplastic producers for their market. The producers of the bio-plastic granules, in turn, depend on the plastic molders to use their material. Greater influence rests with the plastic molders, who are in a position to urge the most influential but probably least interested group of all – the plastic-using public – to choose a bioplastic rather than a conventional plastic product. The stakeholders with less influence, such as the green activists, can participate only by exerting pressure on other stakeholders. The arrows on the figure show the influence paths.

The greater the involvement of stakeholders, the greater is the scope for mutual understanding (Figure 4.2). The stakeholder-analysis diagram forces the questions: Who? What interests? What influence? What interactions? Are the stakeholder concerns justified? If they are, what can be done to accommodate them? If they

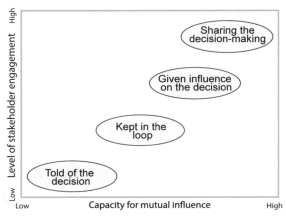

FIGURE 4.2
Levels of stakeholder engagement and influence.

are not, how can they be refuted? To answer those questions we need facts. And that brings us to fact-finding.

4.4   STEP 3: FACT-FINDING

This is the deterministic step: that of assembling factual information about the articulation and the concerns that have been expressed about it. The essence of this step is that it be objective and non-judgmental. The six sectors of Figure 4.3 (here as a reminder) act as prompts. The panels that follow provide check-lists of suggested targets for information gathering. Not all are relevant to all projects. Fact-finding can be researched using readily accessible sources: books, journals, the Internet. Further help is available in the form of the CES EduPack Sustainability database, described in Chapter 15, but it is not essential.

Materials and manufacture. As already said, articulations that make a significant difference are often large in scale and require large quantities of materials. The *bill of materials* (BOMs) lists the quantities. Will the resulting demand for materials, particularly critical materials, put stress on the materials supply-chain? A comparison of demand with supply requires knowledge of the annual world production of a given material. The US Geological Survey[1] provides data for world production and countries of origin, important for ensuring ethical sourcing of materials. Additional information, updated annually,

---

[1]http://minerals.usgs.gov/factbook.

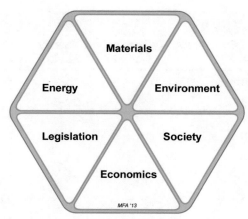

FIGURE 4.3
The six headings for fact-finding. The check-lists suggest key questions.

is available from the United Nations,[2] the World Bank[3] and other agencies, which document the quality of governance, finance, civil rights and environmental policy of the 210 nations of the world. An articulation that depends on materials with unreliable supply-chains is unlikely to be sustainable.

The boxes suggest questions designed to elicit the relevant facts. The answers provide the inputs to Step 4, *synthesis*, described in Section 4.5. They also set gates – go/no-go criteria – that allow articulations that are fatally flawed to be rejected, or at least side-lined, more quickly.

MATERIALS SUPPLY CHAIN

- What is the BOMs for a unit of the proposed articulation?
- Is the material supply-chain secure?
- Are any of the materials listed as "critical"? If so, why?
- Given the scale of the articulation, will the demand for critical materials stress the supply chain?
- Where are the materials sourced? What is the human-rights record of the country of origin?
- How much transport is involved in material sourcing and manufacture?
- Are materials used efficiently? Has light-weighting been addressed? Has packaging been minimized?

---

[2]http://data.un.org/Default.aspx.
[3]http://data.worldbank.org.

Table 4.3   The Energy Intensity of Fuels and Their Carbon Footprints (GWP[b])

Fuel Type	Oil Equivalent kg OE[a]	Energy MJ/liter	Energy MJ/kg	CO_2, kg/liter	CO_2, kg/kg	CO_2, kg/MJ
Coal, lignite	0.45	–	18–22	–	1.6	0.080
Coal, anthracite	0.72	–	30–34	–	2.9	0.088
Crude oil	1.0	38	44	3.1	3.0	0.070
Diesel	1.0	38	44	3.1	3.2	0.071
Gasoline	1.05	35	45	2.9	2.89	0.065
Kerosene	1.0	35	46	3.0	3.0	0.068
Ethanol	0.71	23	31	2.8	2.6	0.083
Liquefied natural gas	1.2	25	55	3.03	3.03	0.055
Hydrogen	2.7	8.5	120	0	0	0

[a]Kilograms oil equivalent (the kg of oil with the same energy content).
[b]Global warming potential.

Energy and power. The *duty-cycle* of a product is the expected intensity of use, from which the energy it consumes can be calculated. Energy carries a carbon footprint that depends on its source. Table 4.3 gives approximate values, useful in assessing environmental impact (coming next).

ENERGY AND POWER

- Does the articulation use energy over its life?
- What is the duty cycle?
- What is the source of power?
- Is energy storage involved? How is the energy stored?
- Have energy-efficiency, life-expectancy and maintenance been considered?

Environment. Many articulations of sustainable development claim to provide functionality with less resource consumption and environmental damage than at present. Establishing whether this is in fact achieved is an important part of the assessment. This requires a life cycle assessment (LCA) or an eco-audit – a fast, approximate life-assessment that is still accurate enough to guide decision-making. Energy use is often the single largest cause of environmental stress – see "Energy and Power", above.

ENVIRONMENT

- What does an eco-audit or LCA of the articulation reveal?
- What is its carbon footprint? Is it less than current practice?
- Does the articulation provide its function with less dissipation of material resources than current practice?
- Can the materials of the product be recycled? Does a recycling infrastructure exist?
- Do the materials, manufacture, use or disposal of the articulation pose any threat to the bio-sphere?

Legislation and regulation. An increasing body of legislation sets requirements for reporting of the use of toxic substances, for documenting energy and material use and for collection and recycling of materials at end of life. Some are guide-lines. Many of these are now legal obligations. An articulation that does not comply with the law is not a sustainable development.

LEGISLATION AND REGULATION

- What legislation or regulatory measures are relevant to the implementation of the articulation?
- What legislation or regulation bears on the production, use and disposal of the materials of the articulation?
- What restrictions do these impose on the sourcing or disposal of the materials required for the articulation?
- Does the articulation comply with any existing or anticipated legislation?
- What legislation, not yet enacted, is anticipated?

Social equity. Governments and corporations increasingly recognize the importance of demonstrating corporate social responsibility and the implied duty to bring benefits to the local communities in which they operate. These include: local investment in education, welfare and health, the use of local labor and respect for local customs, traditions and beliefs. More on this in Chapter 6.

SOCIETY

- Will the articulation, if implemented, create jobs?
- If it creates wealth, will it be shared equitably?
- Are plans in place to invest some of this wealth in the local community?
- Are aspects of manufacture, use or disposal of the product inequitable or exploitative?
- Does it contribute to human well-being, national and personal self-esteem and pride?
- Does it increase self-sufficiency and resilience?
- Does it clash with cultural or societal norms?
- Is the design inclusive, or does it exclude some members of society?

Economics. The cost of implementing an articulation is perhaps the most difficult aspect to estimate – cost over-runs of a factor of two are not uncommon. It is also the most controversial because there are always other projects on which the money could be spent. Establishing the cost-benefit balance of an articulation is an essential part of demonstrating that it is a sustainable development.

ECONOMICS

- What is the cost-benefit balance of the articulation?
- What is the pay-back time?
- What other projects are deprived of resources in order to fund the articulation?
- What subsidies or credits exist to support articulations seen as sustainable?

Time-value of money. Here we need to digress to discuss pay-back and the time-value of money. Money has a time value because of its interest-earning potential.[4] If you invest in a sustainable project (like solar panels on your house) there is an up-front capital cost of, let us say, $10,000. The $10,000 could have been invested in an interest-earning way but now it cannot, because it has gone to pay for the panels. Over time, the panels produce power and generate an income via a feed-in tariff or by reducing your energy bill. The panels have paid

---

[4]For a good introduction see Dieter, G.E. and Schmidt, L.C. (2013) "Engineering Design", 5th edition, McGraw Hill, New York.

for themselves when the total of this income is equal to the up-front cost adjusted for its time value – the interest it could have earned.

What is this interest? And how long is the pay-back time? If a present sum P is invested at a simple interest rate i, the interest each year is iP. After n years the sum will have increased to a value F, its "future worth", of:

$$F = P + niP = P(1 + ni) \tag{4.1}$$

Financial transaction generally use compound interest, not simple interest. The interest is added to the capital at fixed intervals (annually, for instance) and in the next interval interest is paid on the total. Then the future worth in successive years is

$$\begin{aligned}
F_1 &= P + iP = P(1 + i) \\
F_2 &= P(1 + i) + iP(1 + i) = P(1 + i)^2 \\
F_3 &= P(1 + i)^2 + iP(1 + i)^2 = P(1 + i)^3
\end{aligned}$$

So after n years the future worth is

$$F_n = P(1 + i)^n \tag{4.2}$$

Equivalently, a sum of F to be paid in n years' time has a present ("discounted") value of

$$P = \frac{F}{(1 + i)^n} \tag{4.3}$$

The factor $1/(1 + i)^n$ is called the "present value factor", and, when thinking of present value, the quantity i is called the "discount rate" rather than the interest rate.

If the up-front cost were paid off as a single payment at the end of n years, the payment would be that given by Eqn (4.2). But more usually the capital is recovered as a series of equal receipts A, occurring equally at the end of each time period – the annual feed-in tariff for the electricity generated by the panels, for instance – so that the initial capital investment is paid off over time. The recurring payment needed to return the initial capital investment P plus interest on that investment at a rate i over n years is

$$A = P\frac{i(1 + i)^n}{(1 + i)^n - 1} \tag{4.4}$$

This can be solved for n to give the interest-adjusted pay-back time n^* when the annual recovered capital is A, giving

$$n^* = \frac{-\ln\left(1 - \frac{iP}{A}\right)}{\ln(1+i)} \tag{4.5}$$

Spreadsheets like Excel have compound interest factors built in as functions. They include Eqns (4.2)–(4.4).

Example. A 2-kW solar panel-set costs \$4000 when fully installed. The set in practice generates 1600 kWh of electrical power per year. The feed-in tariff is \$0.5 per kWh, generating a return of \$800 per year. How long does it take for the panel-set to pay for itself if the prevailing interest rate 5% per year?

Answer. The pay-back time for the panel-set is

$$n^* = \frac{-\ln\left(1 - \frac{0.05 \times 4000}{800}\right)}{\ln(1 + 0.05)} = 5.8 \text{ years}$$

4.5 STEP 4: INFORMED SYNTHESIS

The *synthesis* step aims to assess the impacts of the information gathered in Steps 1–3 on the three capitals. Put another way, it seeks to evaluate the triple bottom line. We are dealing here with information about different things measured in different ways and which cannot be combined in deterministic ways. Finding a balance requires holistic thinking and informed debate. The outcome is, inevitably, subjective, influenced by social, cultural and political background. The function of the first three steps is to provide a common background of accepted facts on which the informed debate can be based.

Table 4.4 expands the description of the three capitals, with reminders of why each is important, how (very approximately) it can be measured and what can damage it. The synthesis debate centers on how the facts unearthed in Step 3 impact on each of the capitals.

The synthesis-debate can, to a degree, be structured by starting with the check-lists in the panels below.

Table 4.4	Key Features of the Three Capitals			
Capital	**What is it?**	**Why is it Important?**	**What Can Damage it?**	**How do You Measure it?**
Natural capital	Clean atmosphere, fertile land, fresh water, productive oceans, accessible material and energy resources, healthy biosphere	It is the support system for all life, today and in the future	Loss of bio-diversity and habitat, climate change, resource depletion, emissions	Environmental space, carrying capacity, ecological footprint[a]
Manufactured and financial capital	**Manufactured:** built environment, industrial capacity, materials and goods that contribute to the production process **Financial:** a measure of ability to replace or expand manufactured capital	It provides the infrastructure for shelter, food production manufacture, transport, and employment	Recession, breakdown of financial system, armed conflict	GDP[b] GPI[c]
Human and social capital	**Human:** health, knowledge, skills, motivation, happiness **Social:** democracy, freedom of speech, social structures	Satisfaction with life, ability to contribute to society	Inequity of power, wealth, influence	GINI index[d], level of literacy and education Satisfaction with life scale[e]

[a]The ecological footprint is a measure of the productive land area required to support a defined economy or population. Industrialized economies require more land than they own to provide the resources they need. They respond by importing resources from other countries. Dividing the earth's productive land by the number of people today (7 billion) gives an average of 1.9 hectares per person. If some productive land is set aside for other species, the allocation per person becomes smaller. The ranking of national economies by ecological footprint shows which of them are ecologically most sustainable and which are running an ecological deficit.
[b]*Gross domestic product (GDP) is an imperfect measure of manufactured capital, widely used to report the economic state of a nations.*
[c]*Genuine progress indicator (GPI) is considered to be a better economic metric of a country than GDP. The GPI indicator takes everything the GDP uses into account, and adds other figures that represent negative effects related to economic activity (such as the cost of crime, cost of ozone depletion and cost of resource depletion, among others).*
[d]*The GINI index is a measure of the inequality of wealth distribution within an economy. If the wealth of a country was distributed equally, so that everyone got the same, that country has GINI index of 0%. If the opposite were true, so that some people got a great deal and some get none, the GINI index tends towards its ultimate limit of 100% – a situation in which one person gets all the wealth.*
[e]*The* Satisfaction with Life Scale (SWLS) *(Diener et al, 1985) is a subjective assessment of well-being.*

PRIME OBJECTIVE

- Can the prime objective of the articulation be achieved on the scale and in the time envisaged?

NATURAL CAPITAL – PLANET

- Does the articulation reduce dependence on finite resources?
- Does it reduce emissions to air, water and land?
- How are biodiversity and eco-systems affected?
- Does it cause irreversible change?
- Is a rebound effect possible? (Greater efficiency causing increased consumption.)

MANUFACTURED CAPITAL – PROSPERITY

- What will the articulation cost? What revenue will it generate?
- Will it increase industrial capacity?
- How will existing institutions be affected?
- Does it increase employment and livelihood?
- Is it creating new opportunities for development or innovations?

HUMAN CAPITAL – PEOPLE

- How are human health, education and skills affected by the articulation?
- Will the articulation contribute to human happiness and well-being?
- Does it increase knowledge?
- Is it culturally acceptable? Does it affect cultural identity?
- Does it promote equality?
- Is it consistent with principles of freedom of information and speech, good governance and democracy?

STAKEHOLDERS

- Have stakeholders been properly consulted?
- Are stakeholder concerns addressed?
- If they are not, have the concerns been convincingly shown to be without basis?

Table 4.5 is a matrix to help structure thinking and reporting of the synthesis step. On the left are the six sectors of the fact-finding exercise. Across the top are the three capitals with a set of questions to prompt thinking about the effect the facts have on them.

Table 4.5   **The Synthesis Matrix**

		The three Capitals		
		Human capital - People Health? Wellbeing? Convenience? Culture? Tradition? Perceptions? Equality? Morality?	**Natural capital - Planet** *Can prime objective be met? Are stakeholder concerns addressed? Unintended consequences?*	**Manufactured capital - Prosperity** Cost – Benefit? (Cost facts vs. Eco facts) Legitimacy? Conformity with law?
The six sectors	**Materials**	(+) Fact 1 (−) Fact 2		
	Energy	(−) Fact 3	(+) Fact 3	(−) Fact 1 (−) Fact 3
	Environment		(+) Fact 3 (−) Fact 4	
	Legislation	(−) Fact 4		(+) Fact 4 (−) Fact 5
	Economics		(−) Fact 2	(−) Fact 6
	Society	(+) Fact 5 (+) Fact 6	(+) Fact 6	
	Synthesis (the most telling facts)	*Net gain or loss?*	*Net gain or loss?*	*Net gain or loss?*
		The bottom line – food for reflection		

Assemble facts

The idea is to enter the facts themselves into the matrix, attaching a "+" if the impact is broadly positive and a "−" if it is negative, or color-coding them (as here). Not all the cells will have entries; and some facts may appear in more than one column, sometimes contributing in a positive way to one capital and in a negative way to another. When filled as far as is practical, the matrix gives an overview of the overall impact from which broad conclusions can be drawn. It is used in the case studies of Chapters 8 to 13 and can be handed out as a teaching aid.

McDonough and Braungart (2002) suggest a design-tool to avoid over-emphasis on one capital at the expense of others. As the product or system is planned, its consequences are questioned and innovative answers sought by moving on a circular path around the colored triangles of Figure 4.4. At the bottom left corner we are in the realm of pure capitalism. The question: can the project or service make a profit? If the answer is "no", a re-think is needed. A business that cannot stay in business is letting its shareholders down.

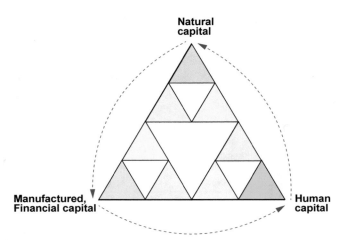

FIGURE 4.4
McDonough and Braungart's tool to explore impact on the three capitals in a balanced way.

Moving a little to the right, questions of fair financial reward arise: Is the workforce paid a fair wage? Is there equal pay for equal work? These are issues of economy tempered by equity. Continuing to the extreme right corner the questions become purely social. Are stakeholders respected? Are their concerns heard in a sympathetic, understanding way? Are actions taken to remedy inequalities, setting aside financial or ecological concerns?

Continuing further round towards the top of the pyramid, environmental issues enter the questioning. Does the project consume resources or release chemicals that might impact human health or future well-being? Is there an acceptable balance between ecology and equity? Reaching the top, questions focus on the environment in the broadest sense. What are the impacts on species other than our own? What effect might the project have biodiversity, habitats and the interdependence plant and animal life?

Moving finally to the third side of the pyramid, money and manufacture re-appear. Can concern for the environment be coupled to economic gain? What adjustments can be made to increase material efficiency, extracting more functionality from less material, and retaining materials in use for longer? How can materials be recovered at the end of life in a more economical way?

4.6 STEP 5: REFLECTION ON ALTERNATIVES

At this point, the facts have been gathered and judgments have been made about their impact on the three capitals. Now is the time to think over the findings. Can the prime objective be implemented on the scale and in the time-frame that is planned? Do its positive contributions to the three capitals outweigh the negative? What are the problems, the unintended consequences or the obstacles?

This is the time for broader thinking. First, the *short term*. Some of the case-studies in later chapters carry many negatives and few positives if implemented in the near-term because of resource-demands or lack of supporting infrastructure. That is not an immediate reason to ditch them; it is better instead to explore the *long-term* picture. Is pain today justified by gain tomorrow? Given the knowledge that has been gathered, is there a more imaginative way to achieve the desired objective? Are there new, possibly disruptive, ways of reaching the same goal? Would they require existing or radically new technology or infrastructure? Could change of human behavior render the objective no longer needed or desirable? This requires methods for creative thinking.

That, of course, is not easy, but creativity tools do exist and can be helpful. Some are summarized in the Appendix of this chapter.

4.7 SUMMARY AND CONCLUSIONS

Assessing an articulation of sustainable development is in some ways like assembling a complex jigsaw puzzle – it is an exercise in finding order in what at first seems like chaos. The many pieces of the puzzle equate to the many and varied pieces of information ("facts") that characterize the articulation. But while the pieces of the jigsaw only fit together in one way, questions of sustainability have more than one answer. The new dimension, absent from the jigsaw, is the need for judgment, compromise and creative thinking.

This chapter provides prompts, check-lists and matrices to help with formulating objectives, stakeholder analysis, fact finding, and the assembly ("synthesis") of these into a coherent picture. Their use is illustrated in the case-studies of Chapters 8 to 13. Experience in using them for teaching is discussed in Chapter 16. Before moving on to the case studies, however, we need to explore two other aspects of materials and sustainable development: that of risk and of corporate sustainability assessment.

Further Reading

Dieter GE, Schmidt LC: *Engineering Design*, 5th edition, New York, 2013, McGraw Hill (A respected introduction to the engineering design process, including planning and economics.).

Freeman RE: *Strategic management: A stakeholder approach*, Boston, Mass, 1984, Pitman Publishing (2010) Cambridge University Press, Cambridge UK, ISBN:978-0-521-15174-0. (One of a number of good introductions to stakeholder analysis.).

Frooman J: Stakeholder influence strategies, *Academy of Management Review* 24(2):191–205, 1999 (One of a number of good introductions to stakeholder analysis.).

How to do Stakeholder Analysis, 2011, http://www.slideshare.net/AberdeenCES/how-to-do-stakeholder-analysis. (One of a number of good introductions to stakeholder analysis.).

McDonough W, Braungart M: *Cradle to Cradle, Remaking the Way We Make Things*, New York, 2002, North Point Press. ISBN:13 978-0-86547-587-8. (The book that popularized thinking about a circular materials economy.).

MindTools Stakeholder Management tool, 2014, http://www.mindtools.com/pages/article/newPPM_07.htm. (Matrices to guide stakeholder analysis.).

Reed MS: Stakeholder participation for environmental management: a literature review, *Biol. Conserv.* 141:2417–2431, 2008 (Stakeholder analysis in an environmental context.).

Reed MS, Graves A, Dandy H, Posthumus H, Hubacek K, Morris J, Prell C, Quinn CH, Stringer LC: Who's in and why: Stakeholder analysis as a prerequisite for sustainable natural resource management, *J. Environ. Manage.* 90: 1933–1949, 2009. Stakeholder analysis in an environmental context.

Stakeholder analysis toolkit: http://www.mmu.ac.uk/bit/docs/Stakeholder-analysis-toolkit-v2.pdf, 2005 (Matrices to guide stakeholder analysis.).

The Stakeholder Engagement Manual, 2005, http://www.accountability.org/images/content/2/0/208.pdf. Stakeholder Engagement Report by Sustainability (One of a number of good introductions to stakeholder analysis.). ISBN 1 901693 220.

4.8   APPENDIX: CREATIVITY AIDS – A BRIEF SURVEY

Introduction. Techniques for creative thinking work by breaking down barriers and forcing new angles of view. Some help the designer to escape from the view of the "product" as it is now, seeking to see it in new ways. Others are ways of extrapolating ideas from the past and present to envisage new solutions for the future. Still others rely on the accumulation of information that relates to a design challenge, slowly building a "scaffold" of ideas that becomes, when the last pieces are in place, the framework of the new design. All require work. Creativity, it is rightly said, is 1% inspiration and 99% perspiration.

Where do good ideas come from? They do not just "happen". Rather, they emerge from a fascination with a problem, an obsession almost, that sensitizes you to any scrap of information that might, somehow, contribute to finding a solution. Combine this with reading and discussion, loading up the mind, so to speak, with background information and with solutions to related problems that, you sense, might be relevant. The human mind is good at rearranging bits of information, seeking patterns, often doing so subconsciously (we have all had the experience of waking in the night with the answer to a problem that, the previous evening, had no solution). It is like finding a route across previously unmapped territory. The route is what is wanted, but to find it you have to map, at least approximately, the territory as a whole. Or (another analogy) it is like building a scaffold out of many scaffold-poles to reach a remote and awkward roof-top. Only when the last pole is in place can you reach the top; till then it was inaccessible. If there is a moment of real creativity it is probably the insight that provided that last pole. But it would have been no help if the rest of the scaffold were not already in place. To repeat: good ideas do not "happen"; they emerge by giving the mind the means to find them.

The subject of materials is one that is rich in unanswered questions. Building a mental library of relevant "bits" of information and cultivating it by allowing reflection-time to try out various arrangements of the bits, allows them to finally fall into a pattern that gives new insight or suggests a new solution.

Reflection is probably the first pre-requisite for creativity. In addition there are techniques, tools and processes. Some people find some work for them, others prefer others. Here is an overview.

Brain-storming, mind maps, sketching, mood boards and inspiration from nature. *Brainstorming* relies on the group dynamics that appear when participants express their ideas, however wild, deferring all value judgment until the process is over. Humor plays a role. A joke that works relies on a creative jump, an unexpected outcome, by-passing normal reasoning. It switches the points, so to speak, deflecting thought off its usual rails onto a new track. It is the creativity of a good joke that gives pleasure, makes you laugh. Introducing it creates an environment for creative thinking. To work, brain-storming sessions must be fun and be kept short – experience suggests that one lasting more than 20 min ceases to be productive.

Mind-mapping can be a sort of personalized brainstorming in which ideas are placed on a page and linked as appropriate; these links are used to stimulate further thinking: "light-weight materials… wood… cells… porous solids… foams… metal foams… titanium foam…?" Or it can be a map of a more considered sort (see below).

Sketching is a creative activity. The form of a new product first takes shape by sketching – freehand drawing, freely annotated, that allows the designer to explore alternatives and record ideas as they occur. Sketching is a kind of visual thinking, an image-based discussion with oneself or with others. Only at a later stage is the design dimensioned, drawn accurately and coded. Modeling software allows the display of projections of the product and experimentation with visual aspects of color and texture. But designers emphasize that it is not these but the act of sketching that stimulates creativity.

A *mood board* is a visual scrap-book, arranged on a large board placed where the designer will see it easily. It takes the form of a personalized, project-focused collection of images, objects or material samples, chosen because they have colors, textures, forms, features or associations that might contribute to the design. Images of products that have features like those sought by the designer and images of the environment or context in which the product will be used act as prompts for creative thinking or using old ideas in new ways.

Nature as inspiration. To "inspire" is to stimulate creative thinking. Inspiration can come from many sources. That for product design comes most obviously from other products and from materials and processes, particularly new ones. After that, nature is perhaps the richest source. The mechanisms of plants or animals – the things they can do and the way they do them – continue to mystify, enlighten and inspire technical design: Velcro, non-slip shoes, suction cups, and even sonar have their origins in the observation of nature. Nature as a stimulus for industrial design is equally powerful: organic shapes, natural finishes, the use of forms that suggest – sometimes vaguely, sometimes specifically – plants and animals: all are ways of creating associations, and the perceived and emotional character of the product.

SCAMPER (van Wulfen, 2013) is an acronym of the first letters of seven approaches that can be used to change a product or service, useful when brainstorming. Here are the seven approaches.

1. S = substitute. What can I replace in the composition, the material, the appearance and the size etc. of the product?
2. C = combine. What can I combine with the product to improve it?
3. A = adapt/adopt. Can I adapt the product to do something new, or adopt features from other sectors to improve this one?
4. M = magnify/minimize/modify? What can I magnify, minimize or modify about the product?
5. P = put to other uses. Can I use the product for something else?
6. E = eliminate. What can I eliminate to improve performance or reduce cost?
7. R = reverse/rearrange. Is there anything I can reverse, turn inside out or do in a different order?

The facilitator recites the seven questions and asks the participants to create new ideas and write them on post-it notes. The post-it notes are surveyed at the end of the exercise.

Case-based reasoning: inductive reasoning and analogy. Inductive reasoning (Kolodner 1993) builds on previous experience. The starting point is a set of requirements expressed as *problem features*. A match is then sought between these and the problem features of other previously solved problems, allowing new, potential solutions ("hypotheses") to be synthesized and tested.

A central feature here is the library of previously solved problems or "cases" – a "case" is a problem, an analysis of its features, a solution and an assessment of the degree of success of this solution.

The challenge in assembling the library is that of appropriate indexing – attaching to each "case" a set of index-words that capture its features. If the index-words are too specific the "case" is only retrieved if an exact match is found; if too abstract, they become meaningless to anyone but the person who did the indexing. Consider, as an example, the "case" of the redesign of an electrical plug to make it easy to grip, insert and pull-out by an elderly person with weak hands. Indexing by "electrical plug" is specific; the case will be retrieved only if the "plug" is specified. Indexing under "design for the elderly" is more abstract, and more useful. Plugs are not the only thing elderly people find hard to use. Cutlery, taps, walking sticks and many other products are adapted for elderly people. Examining their shapes and materials and processes used to make them may suggest new solutions for the plug.

Software shells exist that provide the functionality to create case-based systems, but, for any given domain of problems, the library has to be populated.

Backcasting. Forecasting is extrapolating from the past through where we now are in order to envisage the future. Backcasting is imagining the future and asking what steps are necessary to reach it from where we now are. Forecasting has the tendency to project the problems of today into the future. By starting with a clean slate, backcasting encourages a broader range of thinking and provides a stimulus for creativity. It is seen as a powerful tool to support sustainable development.[5]

It sounds attractive but it is not easy. Getting large groups of people to agree on a desired future scenario is difficult. Scenarios that are too specific can limit innovation and suppress creativity. Proponents[5] argue that strategic sustainable development should be based on "backcasting from a set of sustainability principles which, if violated, could render society un-sustainable". The Canadian non-government organization The Natural Step suggests the four principles shown in Table 4.6.

TRIZ and the 9-windows method. TRIZ (standing for theory of innovative problem solving – but in Russian) is the brainchild of the

---

[5]The Natural Step http://www.naturalstep.ca/about.

Table 4.6	**The Four Sustainability Principles**

1. There must be no systematic increase of concentrations of substances extracted from the earth's crust (for example, heavy metals and fossil fuels)

2. There must be no systematic increase of concentrations of substances produced by society (for example, plastics, dioxins, PCBs and DDT)

3. The systematic physical degradation of nature and natural processes (for example, over harvesting forests, destroying habitat and overfishing) must be eliminated

4. Conditions that systematically undermine people's capacity to meet their basic human needs (for example, unsafe working conditions and not enough pay to live on) must be removed

Russian patent expert Genrich Altschuller, distilled, so to speak, from his study of patents. He arrives at 40 principles and 8 patterns of evolution for creating engineered products. There are disciples of his methods and there are non-believers. Be that as it may, one technique, the 9-windows method claimed as part of the TRIZ tool-set, finds wide use as stimulus for creative thinking. The obstacle to innovation is, often, a preoccupation of the system as it is *now*. The 9-windows method forces a view of the system on different conceptual scales and at times other than the present. It takes the form of a 3×3 matrix of initially empty boxes.[6] The system for which creative thinking is sought is put in the central box – it can be described in words or recorded as a schematic or an image. It represents the system as it is now.

The horizontal axis is time past, present and future (it helps to make this quantitative, thus: 1 year ago, now, one year from now; or 50 years ago, now, and 50 years hence). The vertical axis is that of scale: *subsystem* (the components of the system) in the bottom row; *system*, the scale of the product itself, in the middle row; and *super-system*, the scale of the environment in which the system must operate, in the top row. There are several ways to use the past, present and future columns. One is to ask: what are the antecedents of the system (subsystems, super-system), and what, ideally, would you like these to become in the future? Another is to ask: if you could have designed the system differently (in the past) what would you have done? What can I do now to enhance the system? What – given time – should be the aim for the future? A third is

[6]Some advocates use two 9-windows sets, one to analyze the problem, the other to explore solutions.

to ask: Where are we now? Where have we come from over the last 5 years? Where do we wish to be 5 years from now? The idea is to encourage a view of the problem from in front and behind, and from above and below, allowing a freedom to zoom in and out. The method is sufficiently useful that an example is justified.

Figure 4.5 shows the 9-windows. The product to be re-designed is the electric kettle, shown in the middle box of the "today" column. The kettle has long history. It evolved from kettles that looked like that in the middle box of "yesterday". The aim is a design for the kettle of "tomorrow".

Think first of the super-system in which the kettle will be used. The top row suggest the size, color, décor and equipment of kitchens from yesterday and today, and a guess at how these might look tomorrow – increasingly hi-tech but increasingly small. The re-design must accommodate this evolution – it is the environment in which the kettle-users will live. The ascetic, rectilinear design of the future kitchen suggest a compact, cuboid kettle with similar minimalist aesthetics.

The bottom row reports aspects of the technology of kettles that have changed radically, notably choice of materials and ways of

FIGURE 4.5
The 9-windows method applied to the redesign of an electric kettle.

introducing heat. What opportunities are presented by new materials and heat-transfer methods? Advanced ceramics and glasses are now used in many demanding applications. They have the merit of extreme durability. Heat-transfer technology has changed greatly. It is well to remember, too, the emerging priorities, globally, of conserving energy, mineral resources and water, and the prerequisites of environmental and economic responsibility. How can emerging technology be used to adapt the new kettle to meet these expectations?

The 9-windows analysis broadens the horizons of thinking. It gives doors out of the box in which we started – that of a single focus on today's kettle. Better kettles may not strike you as an elevated goal. Replace it, then, by personal transport (the car) or personal communication (what comes after the smart phone?) or managing your data (hand-written accounts – Excel spread sheets with search – big data clouds with data-mining?).

Post-normal science. Breaking a problem into layers, as in this book, is a way of reasoning. There are others, not incompatible with this one. One, with the weird name "post-normal science", is relevant. Here is the idea.

Problems differ in their level of factual uncertainty and complexity, and in the consequences of getting the answer wrong (Figure 4.6). Standard methods of engineering-science work well when the facts

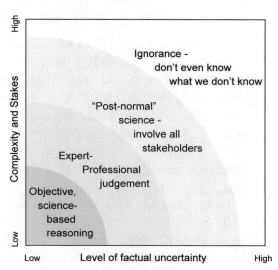

FIGURE 4.6
The levels of uncertainty and methods of tackling them.

are known and the system complexity is limited. An example might be that of analyzing the vibrational response to road roughness of a planned automobile body-shell – current finite element methods handle this well. When the system is more complex and the factual background less complete, seeking the judgment of acknowledged experts and examining current best-practice provides a way forward. An example might be that of designing and commissioning a manufacturing system for the new car – guidance from someone with past experience of such systems is a great help. When facts are even less certain, the complexity is very high and (often) a great deal is at stake, the central issue becomes the agreement of all concerned – the stakeholders, and the wider community – about the course of action. An example might be the decision to invest heavily in the design and production of fuel-cell powered cars. Today both facts and experience are lacking. The key element is that all those that have relevant information bring it to the table and contribute to the debate, assuming mutual responsibility for success or failure. If the consequences of failure are catastrophic in financial, environmental or human terms, the precautionary principle applies: plan to cope with the least-good outcome. That, taken to an extreme, inhibits risk-taking, and most great advances have involved risk, so a balance is required. This last consultative approach carries the name "post-normal science", perhaps unfortunate because it sounds like science by popular vote, but it is treated seriously because it offers a way forward when all others have failed.

The layered method uses all three levels. Stakeholder analysis and fact-finding are objective and systematic, both characteristics of normal science-based reasoning. Synthesis and the final judgment that a proposed articulation is viable almost always involve the resolution of conflict; it is here that experience and informed debate involving all stakeholders is the surest way forward.

Further Reading: Creativity Aids and Reasoning

Altshuller H: *The Art of Inventing (and Suddenly the Inventor Appeared.* translated by Lev Shulyak. Worcester, MA, 1994, Technical Innovation Center, ISBN: 0-9640740-1-X (One of many books about the TRIZ method.).

Burge S: *The systems thinking toolbox.* http://www.burgehugheswalsh.co.uk/uploaded/documents/AF-Tool-Box-V1.0.pdf, 2011 (An introduction to Affinity thinking.).

Funtowicz S: Post-normal Science - Environmental Policy under Conditions of Complexity. http://www.nusap.net/sections.php?op=viewarticle&artid=13. (The approach to complex problem-solving described above.).

Gelb MJ: *How to Think Like Leonardo da Vinci: Seven Steps to Genius Every Day*, NY, 1998, Bantam Dell – Random House, ISBN: 978-0-440-50827-4 (Gelb uses Leonardo's seven "Principles" as a framework for discussing innovative design.).

Innovation. http://www.claytonchristensen.com/disruptive_innovation.html.

Jacoby R, Rodriguez D: Innovation, growth and getting to where you want to go, *Design Management Review*, 2007. Winter 2007. (Jacoby and Rodriguez, Neumeier and Van Wulfen propose ways to stimulate innovation.).

Johnson S: *Where do good ideas come from? The natural history of innovation*, London, UK, 2010, Penguin Books, ISBN: 978-1-846-14051-8. A remarkable survey of innovation across a wide spectrum of industry sectors.

Kelly T, Littman J: *The art of innovation*. NY, 2001, Doubleday, ISBN: 0-385-49984-1. An exposition by the founder of IDEO of the philosophy and methods of his design company.

Kolodner JL: *Case based reasoning*, San Mateo, CA, 1993, Morgan Kaufmann Pub, ISBN: 1-55860-237-2. An introduction to case-based reasoning in design.

Lehrer J: *Imagine: How Creativity Works*, Edinburgh, UK, 2012, Canongate Books. ISBN: 978-1-547-84767-786-0. A set of case-studies of innovation in product design.

Leonard-Barton Dorothy: *Wellsprings of Knowledge: Building and Sustaining the Sources of Innovation*, Cambridge Mass, 1995, Harvard Business School Press.

Neumeier M: *Innovation Toolkit*. Pearson Education (US) New Riders Publishing, 2009. ISBN: 13: 9780321660480. Jacoby and Rodriguez, Neumeier and Van Wulfen propose ways to stimulate innovation.

Rantanen K, Domb E: *Simplified TRIZ: new problem-solving applications for engineers and manufacturing specialists*, USA, 2002, St. Lucie Press, CRC Press, ISBN: 1-57444-323-2. The authors, proponents of the TRIZ method for creating thinking, introduce its use in industry.

Savransky SD: *Engineering of Creativity: Introduction to TRIZ Methodology of Inventive Problem Solving*, USA, 2000, CRC Press. More on the TRIZ method.

Technology Futures Analysis Methods Working Group: Technology futures analysis: toward integration of the field and new methods, *Technological Forecasting & Social Change* 71:287–303, 2004. A comprehensive survey analyzing future technology and its consequences, including technology intelligence, forecasting, road-mapping, assessment, and back-casting.

The Natural Step: http://www.naturalstep.org/en/backcasting, 2014. (An introduction to back-casting.).

Van Wulfen G: *The innovation expedition*. BIS publishers, 2013, ISBN: 978-90-6369-313-8. Jacoby and Rodriguez, Neumeier and Van Wulfen propose ways to stimulate innovation.

4.9 EXERCISES

E4.1 **Prime objectives.** About 30% of the energy consumption in Europe is used to heat and service buildings. European Union Directive 2010/31/EU on the energy performance of buildings was adopted in 2010. It requires that member states ensure that by 2021 all new buildings are the so-called 'nearly zero-energy buildings'. What is the (implicit) prime objective, time scale and size scale?

E4.2 **Prime objectives.** European consumer watchdogs recently released a study on how companies use product obsolescence to stimulate sales. This has provoked a plan currently before the French Senate[7] which would, if enacted, extend the legal minimum life of products from the present two years to five years by January 1, 2016. Express this proposed legislation articulation of sustainable development as a prime objective with a size and time scale.

E4.3 **Prime objectives.** The British government plans to introduce a 5p charge for plastic bags in England in a bid to discourage their use. Ed Davey, Secretary of State for Energy and Climate Change, said the charge reflected concern at the environmental impact of the bags, particularly on water-borne animals. The charge, which will only apply to supermarkets and larger stores, begins in 2015. "Avoid this charge. Re-use your plastic bags" he said. (BBC report, September 14, 2013). Express this proposed legislation articulation of sustainable development as a prime objective with a size and time scale.

E4.4 **Stakeholders.** The proposed charge on plastic bags described in Exercise 4.3 will not be without opposition. Who are the stakeholders? What are their concerns?

E4.5 **Fact-finding: material supply-chain.** A leading maker of loudspeakers for high-end audio systems uses strontium ferrite, $SrO \cdot 6Fe_2O_3$, for the magnets of the speakers. The CEO is concerned about the security of supply of strontium. What else is strontium used for? Which nations produce

---

[7]See more at: http://www.connexionfrance.com/Planned-obsolescence-obsolete-products-iPod-washing-machine-printers-view-article.html#sthash.GDAGwgef.dpuf.

strontium? Does the list give any cause for concern? (Producing nations and uses for minerals are found in the USGS annual minerals publication.[8]) Does strontium appear on the list of critical materials list of Chapter 1?

E4.6 **Fact-finding: energy.** Most nations seek to increase production and reliance on renewable energy (hydro, wind, solar PV, and bio-based). All, to some degree, are seasonal and subject to "good" and "bad" years, like agriculture – they cannot just be switched on. This intermittent production creates a growing need for grid-scale energy storage. Use the Internet to research systems for storing energy equivalent to 5% of national power needs.

E4.7 **Fact-finding: legislation.** EU legislation requires the recycling of 85% by weight of vehicles at end of life. The lithium-ion battery of a Nissan Leaf (a plug-in electric car) weighs 272 kg and accounts for over 20% of the vehicle weight. What happens to it at end of life?

E4.8 **Fact-finding: society.** Failure to get the economics right can torpedo an articulation. Cost overruns of a factor of 2 are not unusual; factors of 10 happen, putting a big minus in the "manufactured and financial capital" box. As an example, the final cost of the Sydney Opera House was more than 10 times over budget and it is extravagant in its use of materials – but does this mean it should be written-off as an articulation that is not a sustainable development? What about its contribution to human capital? Debate.

E4.9 **Fact-finding: economics.** You have borrowed $10,000 to install solar panels on your house. If you repay the loan after 10 years at a compound interest rate of 8% per annum, what is the lump sum you have to pay back at the end of the loan period?

E4.10 **Fact-finding: economics.** You have borrowed $10,000 to install solar panels on your house but instead of paying it off as a lump sum after 10 years you choose to make 10 equal payments, one per year, such that the loan is full paid after 10 years. What will these payments be?

---

[8]http://minerals.usgs.gov/minerals/pubs/commodity/myb/.

E4.11 **Fact-finding: economics.** A commercial scale 2-MW wind turbine costs €2.8 million, installed. The feed-in tariff for turbines of this size is at present €0.15 per kWh.
   a. If the capacity factor (the fraction of the rated power that the turbine actually delivers, averaged over time) is 0.2 and the discount rate is 5%, how long will it take the turbine to pay for itself?
   b. At what discount rate would the turbine *never* pay for itself?

CHAPTER 5

Materials Supply-Chain Risk

Chapter Outline

5.1 Introduction and Synopsis 85
5.2 Emerging Constraint on Material
Sourcing and Usage 86
5.3 Price Volatility Risk 88
5.4 Monopoly of Supply and
Geopolitical Risk 89
5.5 Conflict Risk 91
5.6 Legislation and Regulation
Risk 92

5.7 Abundance Risk 94
5.8 Changing Expectation of
Corporate Responsibility 95
5.9 Managing Risk 96
5.10 Summary and Conclusions 97
5.11 Exercises 97
Further Reading 100

5.1 INTRODUCTION AND SYNOPSIS

Two further topics are related to questions of sustainable development. The first is that of supply-chain security and the risk of disruption of material markets. The second is that of corporate sustainability assessment and reporting. This chapter and the next describe the issues that are involved. We can then move on to the case studies.

Materials and Sustainable Development. http://dx.doi.org/10.1016/B978-0-08-100176-9.00005-0
Copyright © 2016 Elsevier Ltd. All rights reserved.

Even a sustainable economy needs materials. A century ago, developed nations sourced materials locally or from colonies or dependencies under their control. Today, materials are sourced and traded globally between independent nations, many with their own developed or developing manufacturing base. This change brings with it supply-chain risks that escalate as affluence and demand grow. As we saw in Chapter 1, these concerns prompt nations to classify risk-prone materials with high commercial or strategic importance as "critical". Many of these critical materials come from one or a very few source-nations, exposing their supply-chain to further restriction for geo-economic or geo-political reasons. These supply-chain risks cause concern at the corporate as well as the national level.

Sustainability in a material-dependent industry, from a corporate viewpoint, means assuring reliable access to affordable raw materials. A recent survey[1] of the CEOs of manufacturing companies in Europe found that 80% of them saw future raw material shortages and rising material prices as risks to their business. Similar surveys highlight the concerns created by the ballooning body of legislation that restricts or bans, for health or environmental reasons, an increasing range of materials and chemicals. Further legislation sets new standards for ethical material sourcing, product safety, consumer protection and responsibility for materials at end of product life. Complying with this legislation can be burdensome and expensive, yet failure to comply can mean loss of business[2]. Thus the competencies of a future engineer have to include not only technical skills but also the ability to manage risk.

5.2 EMERGING CONSTRAINT ON MATERIAL SOURCING AND USAGE

Over much of the last century material supply was not (with occasional exceptions) a major issue. Trade tended to be national rather than global. Material prices, in real terms, were static or falling

---

[1]Government Office for Science UK (2013) "Future of manufacturing: a new era of opportunity and challenge for the UK – summary report", London, available on-line: https://www.gov.uk/government/publications/future-of-manufacturing/future-of-manufacturing-a-new-era-of-opportunity-and-challenge-for-the-uk-summary-report.
[2]Granta Design (2013) "Enabling product design and development in the context of environmental regulations and objectives" a White Paper, available on-line: http://www.grantadesign.com/emit/.

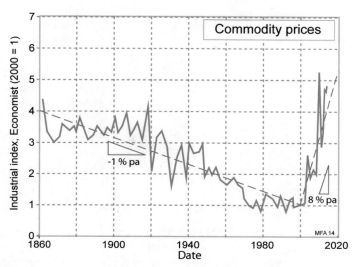

FIGURE 5.1
The movement of price of a portfolio of materials. Data from the World Bank, 2013 and the Economist, 2011.

(Figure 5.1, left hand side). There was relatively little control over the way materials were used or what happened to them at the end of product life.

Today the picture looks rather different[3] (the steep upward slope on the right of Figure 5.1). Since 2000 material prices have risen much faster than inflation. As we saw in Chapter 1, increasing complexity of products creates a dependence on a larger number of elements, some of which are comparatively rare and localized. Manufacturing nations increasingly compete for exclusive rights to minerals resources world-wide in order to safeguard their industrial capacity.

Sustainable corporate development requires anticipating and dealing with risk (Figure 5.2). Material-dependent companies are particularly exposed to risk from supply-chain disruption or breakdown of global markets, from legislation that renders their products uncompetitive or illegal and from changing perceptions of company responsibilities. Risk can be managed by anticipating where change will occur and planning ways to deal with it. Managing risk of this sort is part of corporate sustainability strategy. In the following sections we look briefly at supply-risks associated

[3]http://blogs.worldbank.org/prospects/category/tags/historical-commodity-prices.

FIGURE 5.2
Emerging material constraints.

with materials. Many of these risks feature in the Business section of today's newspapers and news reports. We shall draw on these to illustrate the concerns.

5.3   PRICE VOLATILITY RISK

Headline: Copper prices jump after strike in Chile...

The Sunday Times, July 31, 2011

Headline: Mine strike to drive up platinum price...

Eyewitness News, March 6, 2014

Materials are sourced and traded globally. The ores and feedstock needed to make some of them are widely available, allowing a free market to operate smoothly. For others, however, market forces work less well. The result is disruptive price fluctuations, some of them dramatic. The price of copper (Figure 5.3(a)), for instance, rose by a factor of 4 in 2007 and again in 2011, caused in part by strikes in the copper mines of Chile and a surge in demand from China. The supply of cobalt was disrupted in 2007 by war in the Democratic Republic of the Congo (Figure 5.3(b)), causing the price to increase three-fold. The price of rare earth elements (Figure 5.3(c)) rose by more than a factor of 10 when China imposed export restrictions in 2011. The price of nickel is expected to rise after Indonesia banned exports of metal ores in 2014.

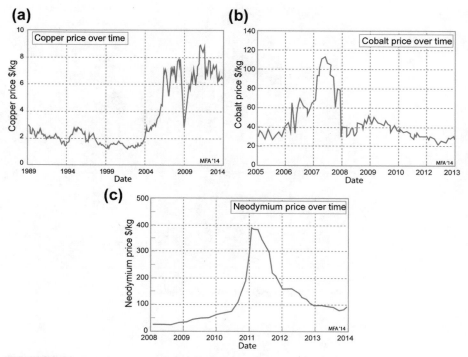

FIGURE 5.3
Fluctuations in material prices (a) copper, (b) cobalt and (c) neodymium.

5.4   MONOPOLY OF SUPPLY AND GEOPOLITICAL RISK

Headline: China still bans rare earths to Japan

The New York Times, November 11, 2010

Headline: Indonesia ban on raw mineral exports threatens nickel shake-up

Reuters, January 10, 2014

As we saw, risk-prone materials with high commercial or strategic importance are classified as "critical". Where do these critical materials come from? And how do we quantify the risk?

Figure 5.4 identifies the main countries-of-origin of the critical materials listed in Chapter 1. If one of these controls a large fraction (a half, say) of global supply of a material, the market in that

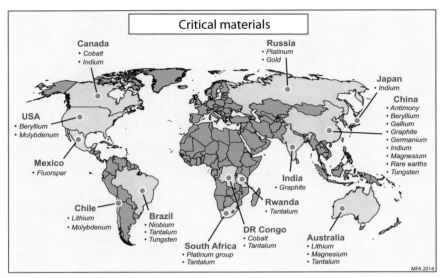

FIGURE 5.4
Predominant countries of origin of critical materials.

material is unbalanced and the potential for disruption exists. The Herfindahl-Hirschman Index (HHI) is a measure of the risk when the supply of a material is controlled by one or a very few nations. This is the index:

$$HHI = \sum_{i=1}^{n} f_i^2 \qquad (5.1)$$

where f_i is the fraction of the market sourced from nation i and n is the total number of source-nations.

The value of the HHI ranges from 0 to 1. If one nation has a complete monopoly of the market, the HHI = 1. If two have equal shares, the HHI = $0.5^2 + 0.5^2 = 0.5$. If the market is served by a very large number of source-nations all of which have only a small fraction, the HHI tends towards 0. An HHI value below 0.1 is an indicator of an unconstrained market. An HHI value above 0.25 indicates severe supply-chain concentration. Figure 5.5 shows the HHI for 20 elements that are classified as critical – most of them have an HHI index exceeding 0.3. Dependence on materials with supply-chain concentration carries risk of monopoly action, disruption of supply and volatile pricing.

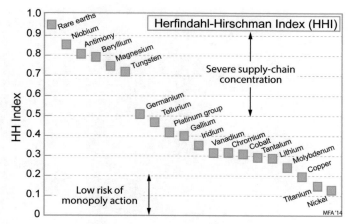

FIGURE 5.5
The Herfindahl-Hirschman Index (HHI) for a number of critical materials. Data from USGS, 2012.

Example: restricted access to raw materials[4]. A number of nations are at present in dispute with China over free trade in raw materials. The dispute concerns certain measures imposed by China affecting the exportation of certain forms of bauxite, coke, fluorspar, magnesium, manganese, silicon carbide, silicon metal, yellow phosphorous, and zinc. The United States, Mexico, and the European Union (the "complainants") challenge four types of export restraints imposed on the different raw materials at issue: (1) export duties; (2) export quotas; (3) minimum export price requirements; and (4) export licensing requirements.

[4]European Union Enterprise and Industry (2014) "Defining 'critical' raw materials" http://ec.europa.eu/enterprise/policies/raw-materials/critical/index_en.htm.

5.5 CONFLICT RISK

Headline: Apple plans to cut the "conflict minerals" in its devices

The Times, February 14, 2014

Headline: Conflict minerals rule poses compliance challenge

Journal of Accountancy, April, 2013

Headline: Auto industry steels itself for conflict materials rule

Automotive News, September 24, 2013

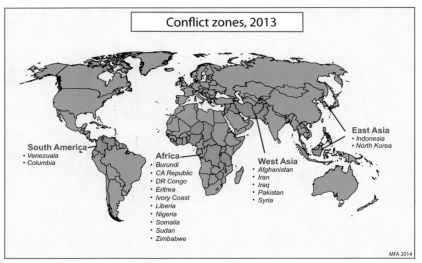

FIGURE 5.6
Nations with on-going conflicts in 2013

Political instability or armed conflict may interrupt supply completely – we have already mentioned the disruption of cobalt supply by war in the D.R. Congo. Increasingly, ethical concerns inhibit the purchase of minerals from nations in which the proceeds might be used to fund conflict. The US Dodd-Frank act, for instance, creates trade barriers with nations in which ongoing conflict compromises human rights. Many of these troubled nations are mineral-rich (Figure 5.6). The short-term loss of production drives up prices. This in turn stimulates expansion of production in other nations and the search for substitutes, ultimately restoring the balance – but this rebalancing typically takes three years.

5.6 LEGISLATION AND REGULATION RISK

Headline: To universal surprise, the National Assembly (of France) bans phthalates...

Le Figaro, May 4, 2011

Headline: Manufacturers unaware of chemical legislation REACH...

BusinessGreen, January 14, 2011

An expanding body of legislation, regulation and standards address the impact of products and the materials they contain on the environment, human health and society. They affect the entire product lifecycle, from substances used in manufacturing to disposal at the end of life. Table 5.1 gives examples covering the reporting and restriction of the use of certain materials, energy usage, conflict minerals, and product disposal at end-of-life. Such legislation creates a considerable burden of compliance and reporting. It also creates risks by rendering key materials or processes obsolete or unobtainable. For long-lived products such as aircraft there is a risk that the materials it contains may become restricted, requiring that substitutes be found and re-qualification be sought – an expensive imposition.

To comply with this legislation, manufacturers must report the chemical background of the materials they use. Ordinary solder contains

Table 5.1	Examples of Legislation that Affect the Use of Materials, Energy and Water	
Sector	**Legislation**
Registration and control of hazardous materials and chemicals	EU Registration Evaluation & Authorization of Chemicals Directive (REACH)
	EU Restriction of Hazardous Substances Directive (RoHS)
	EU Volatile Organic Compounds Directive (VOC)
	US Toxic Substances Control Act (TSCA)
	California Green Chemistry Initiative
	Norwegian Restriction of Hazardous Substances (RoHS)
	China REACH
	China Restriction of Hazardous Substances (RoHS)
Ethical material sourcing	US Frank-Dodd Act
Energy and product design	EU Energy-using Products Directive (EuP)
	EU Energy-efficient Building Directive (EEB)
	France Grenelle 2 Regulations
Water usage	EU Water Framework Directive
End-of-life and control of waste	EU Waste Electrical and Electronic Equipment Directive (WEEE)
	EU End of Life Vehicles (ELV) Directive
	EU Batteries Directive
	Japanese Household Appliance Recycling Law (HARL)

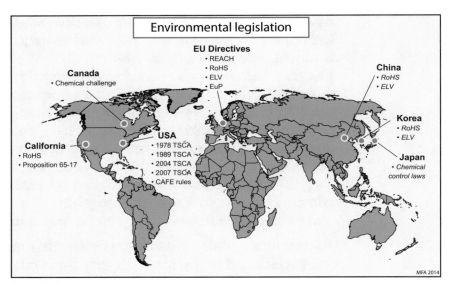

FIGURE 5.7
Some of the legislation bearing on materials enacted since 1976. The titles are defined in Table 5.1.

lead. Rechargeable batteries contain cadmium and long life, low drain batteries contain mercury. All are unacceptable "heavy" metals that damage bio-systems. Chromium plating involves toxic hexavalent chromium. Many polymers contain flame retardants or plasticizers, some recently banned, others on a waiting-list. Many manufacturing processes involve the use of organic solvents. Increasingly, manufacturers must also take responsibility for materials at the end of product life. Legislation is already in place requiring manufacturers to take back electronic equipment, vehicles and tires.

Restrictions can cross national borders. Thus a Japanese car-maker must comply with European or US legislation if they wish to sell their cars in those nations. In an era of globalization, national legislation has global implications. Figure 5.7 dramatizes the rise in legislation since 1976.

5.7 ABUNDANCE RISK

Headline: Are global lithium deposits enough to meet electric car demand?

The New York Times, July 28, 2011

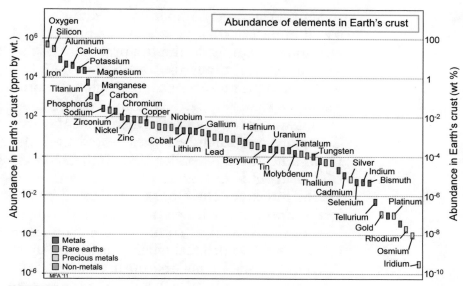

FIGURE 5.8
The abundance of elements in the earth's crust.

How much of everything have we got? Figure 5.8 shows the abundance of the elements in the earth's crust. The range is enormous – a factor of 10^{10} from the lowest (Iridium) to the highest (oxygen). The lowest economically viable concentration for extraction follows a similar pattern. The price follows an inverse picture.

Views are sometimes expressed that we are about to "run out" of some of the less-common elements. That does not look like a near-term problem – the earth's crust is large and deep. But there are, none the less, problems with a material that occurs only in very low concentrations. There is the energy it takes to expose, mine and crush the rock in which it lies. The richest iron ores contain almost half a kilogram of iron per kilogram of ore, but to get 1kg of platinum requires the mining of about 500,000kg of ore. If energy prices rise, the less abundant elements are the most vulnerable.

5.8   CHANGING EXPECTATION OF CORPORATE RESPONSIBILITY

The Dodd-Frank act, mentioned earlier, is an example of the imposition of ethical standards by government. Increasingly consumers, often supported by share-holders and non-governmental organizations, urge companies to examine the environmental and social impact of their processes and products. The United Nations encourages sign-up to the ten principles of its Global Compact[5], a set of ethical standards described more fully in Chapter 6. The vision is to promote a sustainable global economy in which organizations manage their economic, environmental, social performance and governance in ways consistent with the common good. Company performance is monitored by independent organizations that award "sustainability indices" rankings that are now of considerable influence. Prominent among them are the FTSE-4Good index, the Dow Jones Sustainability Index and the Business in the Community Index (FEE 2008). Strong performance in this area can be turned to competitive advantage, but environmental claims made on product labels must stand up to scrutiny. Signs of "Green-washing" – making unsubstantiated claims of good environmental and ethical behavior – are now examined closely; the US Federal Trade Commission[6], for example, sets guidelines for making environmental claims. Rigor and quantitative validation are now essential in this area.

All of these issues can result in significant costs or risks to a business. But considering environmental objectives and regulations during product design and development is no longer an 'optional extra', it is a business imperative.

5.9   MANAGING RISK

Knowing that a material on which you depend is listed as *critical* flags a risk. If you are going to address the risk you also need to know why it is critical and what can be done about it. There are a number of options.

- Short-term fluctuations in availability and price can be damped by building reserves or by negotiating long-term supply contracts.
- Longer-term supply constraints can be mitigated by redesign to use critical materials more efficiently and by increasing take-back, recycling and reuse to keep them in service.

---

[5]UN Global Compact, available on-line: www.unglobalcompact.org/cop/index.html.
[6]US Federal Trade Commission. *Guides for the use of environmental marketing claims*. Part 260.

- Damaging price inflation can be countered by seeking substitutes, and here there is scope for forward-planning to build a knowledge-base of viable substitutes and their suppliers.
- Legislation that creates unnecessary barriers to the sourcing and use of materials can be confronted with reasoned argument and lobbying.

5.10 SUMMARY AND CONCLUSIONS

Growing population and global affluence drive up demand for materials and give mineral-rich nations greater control of availability and price. Unstable government and armed conflict can interrupt supply. Environmental legislation restricts the use and disposal of an increasing number of materials. Pressure for ethical corporate behavior further limits available sources. The combination of these has caused material prices to rise faster than wealth for the first protracted period in the last 100 years.

The role of the Materials Engineer in the twenty-first century is now likely to involve issues such as

- Anticipating material supply-chain constraints and their cause and probable duration, particularly where "critical" materials are involved,
- Precautionary exploration of substitutes for materials important to the enterprise for which they work,
- Adapting to, and complying with environmental and other material-related legislation, and
- Helping the enterprise to adapt to a more circular materials economy, retaining full-life ownership of the materials of their products, maximising reuse and recycling.

5.11 EXERCISES

Tungsten-producing Nation	Tonnes/year 2011
P.R. China	60,000
Russia	3,100
Canada	2,000
Portugal	1,300
Bolivia	1,200

Tungsten-producing Nation	Tonnes/year 2011
Austria	1,100
Other countries	3,400
World	**72,000**

Minerals.usgs.gov/minerals/pubs/commodity

E5.1   The HHI is a measure of risk when the supply of a material is controlled by one or a very few nations. It is defined in Eqn (5.1) of the text as

$$HHI = \sum_{i=1}^{n} f_i^2$$

where f_i is the fraction of the market provided by nation i and n is the number of nations.

The table lists the world production of tungsten by nation. Calculate the HHI index for tungsten based on these values. Does the result suggest supply-chain constraint?

Manganese-producing Nation	Tonnes/year 2011
South Africa	3,400
China	2,800
Gabon	1,500
India	1,100
Brazil	1,000
Ukraine	340
Mexico	170
Other countries	1,400
World	**14,000**

Minerals.usgs.gov/minerals/pubs/commodity

E5.2   The HHI is defined in Exercise E5.1. The table lists the world production of manganese by nation. Calculate the HHI index for manganese based on these values. Does the result suggest supply-chain constraint?

E5.3   **Sourcing cobalt.** You are consulted by a steel-maker who wishes to make a special steel with 5% cobalt content. What can you tell them about the risk to supply of cobalt? Use the USGS Minerals Commodity Yearbook (http://minerals.usgs.gov/minerals/pubs/commodity/myb/) to research the nations from which cobalt is sourced. Use the Internet to research further the stability and governance of the main producing nation (www.govindicators.org). Include the HHI in your report.

E5.4   A company specializing in road-repair relies on cobalt bonded tungsten-carbide cutting tools for stripping road surfaces. The CEO of the company is considering investment in a supplier of these cutting tools to ensure supply. Both cobalt and tungsten appear on the European Union list of critical materials. The HHI for tungsten is 0.67; that for cobalt is 0.31. Both values are high, tungsten particularly so, indicating severe supply-chain concentration. This carries the risk of price fluctuations, confirmed in the case of cobalt by Figure 5.3(b). A search reveals that tungsten price has been less volatile; it has risen by a factor of 1.5 between 2010 and the present day. The CEO of the company is concerned about this and asks for a study of the risk that investment in the company might carry. Investigate what substitutes could be found if the supply cobalt or tungsten was restricted.

E5.5   Germanium is listed as a critical material. Use the USGS Minerals Commodity Yearbook (http://minerals.usgs.gov/minerals/pubs/commodity/myb/) to research the nations from which germanium is sourced. Investigate what germanium is used for, why it is classified as critical and what substitutes could be found if germanium supply was restricted.

E5.6   Platinum is listed as a critical material. Investigate why it is given this classification and what substitutes could be found if platinum supply was restricted.

E5.7   **Manganese** is listed as a critical material. Investigate why it is given this classification and what substitutes could be found if manganese supply was restricted.

E5.8   Rhodium is listed as a critical material. Investigate why it is given this classification and what substitutes could be found if rhodium supply was restricted.

Further Reading

Dow Jones Sustainability Index: . (A ranking of Corporate performance based on sustainability factors that are financially relevant for businesses) http://www.sustainability-index.com/, 2012.

Ethical Corporation: *Full Product Transparency is the Future of Reporting*, (Standards for reporting ethical manufacture and sourcing) http://www.ethicalcorp.com/communications-reporting/full-product-transparency-future-reporting, 2013.

European Union Enterprise and Industry: *Defining 'Critical' Raw Materials*, (A European perspective on critical materials) http://ec.europa.eu/enterprise/policies/raw-materials/critical/index_en.htm, 2014.

Jaffe R, Price J: *Critical Elements for New Energy Technologies*, USA, 2010, American Physical Society Panel on Public Affairs (POPA) study, American Physical Soc (Critical materials for energy technology).

Mulder K, editor: *Sustainable Development for Engineers: A Handbook and Resource Guide*, Sheffield UK, 2006, Greenleaf Publishing Ltd. (ISBN-13: 978-1-874719-19-9. (A set of essays the aims to give engineering student insight into the challenge of sustainable development, the potential contributions of engineering and the barriers and pitfalls).

Rankin WJ: *Minerals, Metals and Sustainability – Meeting Future Material Needs*, CSIRO Press, 2011, ISBN: 9-7806-4309726-1(A thorough and revealing analysis of materials, sustainable development and the Australian mining industry.).

USGS Circular 2112, Wagner LW: *Materials in the Economy – Material Flows, Scarcity and the Environment*, US Department of the Interior, 2002 (A readable and perceptive summary of the operation of the material supply-chain, the risks to which it is exposed, and the environmental consequences of material production) www.usgs.gov.

World GDP per capita over time: https://www.cia.gov/library/publications/the-world-factbook/geos/xx.html (A striking animation of the way in which the GDP of nations has evolved).

US Department of Energy: *Critical materials strategy*, Office of Policy and International Affairs, 2010. materialstrategy@hq.doe.gov, www.energy.gov(A study of materials critical for the energy, communication and defense industries, and the priorities for securing adequate supply).

CHAPTER 6

Corporate Sustainability and Materials

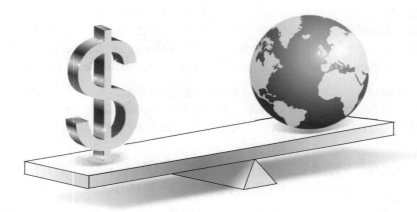

Chapter Outline

6.1 Introduction and Synopsis 101
6.2 Corporate Social Responsibility
    and Sustainability Reporting 102
6.3 Case Studies: Corporate
    SRs 105
6.4 Summary and Conclusions 108
6.5 Exercises 108
Further Reading 109

6.1 INTRODUCTION AND SYNOPSIS

Until recently, the strategic planning of many materials-intensive Corporations and Companies prioritized the financial returns to their owners and shareholders. Narrow focus on financial return alone may bring success in the short term. But there is now an awareness that the longer-term financial success of a company depends on securing the resources in a broader sense: assured access to raw materials, human skills, a marketplace for its products and a reputation that supports and expands its market share. Success, in

Materials and Sustainable Development. http://dx.doi.org/10.1016/B978-0-08-100176-9.00006-2
Copyright © 2016 Elsevier Ltd. All rights reserved.

this view, is measured not only in financial terms but also in terms of stewardship of the resources on which the company draws, the welfare of its employees and the health of the local economy of the regions in which the company does business.

It should be remembered that companies are not people. The primary responsibility of a public company remains that of maximizing return to shareholders while acting within the law – but this is not incompatible with the broader view just described. Increasingly, Corporations and Companies see their social and environmental responsibilities not as "optional extras" but as business imperatives able to create competitive advantage and build respect, trust and loyalty.

This chapter explores these challenges and responses to them. We are in the world of business-speak here, so there are a lot of acronyms. They start in the heading of the next section.

6.2   CORPORATE SOCIAL RESPONSIBILITY AND SUSTAINABILITY REPORTING[1]

Corporate social responsibility. In the past, the primary responsibility of a corporation was to provide an acceptable return on investment to its owners and stockholders. Since 1960s, a broader view of corporate responsibilities has emerged that includes not only stockholders but other stakeholders too: employees, suppliers, customers, the local community, local, state, and national governments, environmental groups and other special interest groups. From this developed the concept of *corporate social responsibility (CSR)*, meaning the expectations that society has of a corporation or company. The underlying assumption is that companies have moral, ethical, and philanthropic responsibilities in addition to their responsibilities to earn a fair return for investors and to comply with the law.

More specifically, the expectations are as follows:

- *Economic:* that Corporations will act with the shareholder interests in mind.
- *Legal:* that Corporations will comply with the laws that govern competition in the market including consumer protection and product laws, environmental laws, and employment laws.

---

[1]A good review of corporate social responsibility is given by www.referenceforbusiness.com/management/Comp-De/Corporate-Social-Responsibility.html.

- *Ethical:* that Corporations will meet societal expectations that go beyond the law by conducting their affairs in a fair and just way.
- *Discretionary:* that Corporations will meet society's expectation to act as good citizens via philanthropic support of programs benefiting a community, or donating employee expertise, time and financial support to worthy causes.

Sustainability reporting. A *sustainability report (SR)* is a way for a company to report the economic, environmental and social aspects of what it does and the way it governs itself. The first such reports were published in the 1980s as a way of countering the negative image that some chemical and mining companies had acquired by their treatment of the environment and of the local communities in which they operated.

Today, Sustainability reporting has become a tool for managing change towards sustainable business practice that combines long-term profitability with ethical behavior, social justice and environmental care. If done responsibly, an SR can enhance brand image and company reputation. It can catalyse change within the company, it helps set new goals and it increases transparency. But it can also be misused to suggest that a company has an environmental and ethical agenda when in reality little is done – a practice known as "green-washing".

Sustainability reporting is not, at present, mandatory but there are discussions at the EU level to make it so[2]. The Global Reporting Initiative[3] (GRI) provides a framework for doing this, based on the ten Principles of the UN Global Compact, listed in the box at the end of this section. Participants commit to the Principles and to publishing an annual Communication on Progress. The GRI encourages reporting under three categories as follows:

- **Economic:** financial results, market presence and procurement practice
- **Environmental:** use of materials, water and energy; emissions; and compliance with environmental legislation
- **Social:** labor practice, human rights, society, product responsibility – broadly, adherence to the UN Global Contract.

---

[2]http://ec.europa.eu/environment/resource_efficiency/news/up-to-date_news/03062013_en.htm.
[3]http://www.globalreporting.org.

The vision is to promote a sustainable global economy in which organizations manage their economic, environmental, social performance and governance in a responsible way. Many Corporations now use it.

Company performance in sustainability reporting is monitored by independent organizations that award "sustainability index" rankings. They are now of considerable influence and are used as guides for environmentally and socially responsible financial investment. The leading indices are summarised below. Most require a subscription for access; one – the CR Index – provides open access.

- **The FTSE4Good Index**[4] claims to be an objective measure of the performance of companies in meeting globally recognised corporate responsibility (CR) standards.
- **The Dow Jones Sustainability World Index**[5] identifies global sustainability leaders through their Corporate Sustainability Reporting. It is based on a consistent improvement in their strategies for climate change, energy consumption, human resources development, knowledge management, stakeholder relations and corporate governance.
- **The Community's CR Index**[6] ranks companies on their environmental stewardship, their social contributions and the extent to which responsible business practice is integrated into their strategy.

UN GLOBAL COMPACT[7] – THE TEN PRINCIPLES

Human Rights – Businesses Should
1. Support internationally proclaimed human rights.
2. Make sure that they are not complicit in human rights abuses.

Labour – Businesses Should
3. Uphold the freedom of the right to collective bargaining.
4. Eliminate all forms of forced and compulsory labour.
5. Eliminate the use of child labour.
6. Eliminate discrimination in hiring and promotion.

---

[4]FTSE4Good Index, 2014. http://www.ftse.co.uk/Indices/FTSE4Good_Index_Series/index.jsp.
[5]Dow Jones Sustainability World Index, 2014. http://djindexes.com/sustainability/?go=literature and http://www.sustainability-indices.com/.
[6]Community's CR Index, 2014. http://www.bitc.org.uk/our-services/benchmarking/cr-index.

Environment – Businesses Should

7. Support a precautionary approach to environmental challenges.
8. Undertake initiatives to promote greater environmental responsibility.
9. Encourage the development and diffusion of environmentally friendly technologies.

Anti-Corruption – Businesses Should

10. Work against corruption in all its forms.

[7]www.unglobalcompact.org/cop/index.html.

6.3   CASE STUDIES: CORPORATE SRs

Many companies now issue SRs. They vary greatly in content and what might be called attitude: some paint a golden picture; others are more direct about their present shortcomings and aspirations. Ploeg and Vanclay[8] suggest criteria by which a SR might be judged. Here is a simplified version.

- Is the report clear, concise and intelligible to the relevant stakeholders?
- Does it use an established reporting framework, such as the GRI?
- Does it explain how the company incorporates CSR into its internal organizational strategy and that of its suppliers?
- Does it identify the stakeholders and their expectations and concerns, and the company's response to them?
- Does it provide adequate evidence (e.g. data) to support the claims that it makes? Are there signs of "green-washing"?
- Does the report establish its credibility, for example through independent assurance by the FTSE, the DowJones or the CR Index of Corporate Sustainability?

Here are brief summaries of SRs from two companies, both with an international reach. The first is a material supplier; the second a material user. Both reports were highly ranked by leading Sustainable Index organisations. The striking difference between the two companies is the nature of the legal and societal pressures to which they are exposed. The company's reaction to these pressures dominates their SR.

[8]Van der Ploeg, L., Vanclay, F., 2013. Credible claim or corporate spin?: A checklist to evaluate corporate Sustainability reports. *Journal of Environmental Assessment Policy and Management*, 15. http://www.worldscientific.com/doi/abs/10.1142/S1464333213500129.

CASE STUDY 1: A MATERIAL SUPPLIER

AngloAmerican[9] is a large mining company with operations in North and South America, Africa and Australia producing iron, copper, manganese, nickel, platinum and diamonds. Mining, in the past, has incurred a negative image of exploitation and environmental negligence. The Chairman's opening statement recognises that the role of big business in society is now under close scrutiny and that greater transparency and accountability are needed to demonstrate commitment to being a responsible corporate citizen.

The mining business, like the oil business, has certain defining characteristics.

- It involves large-scale projects that require significant investments over long-time periods.
- The resources that are mined are sometimes in remote, environmentally sensitive and technically demanding locations, possibly in countries with uncertain political and regulatory situations.
- The resources themselves belong to sovereign states; the company's licence to operate is based on mutual benefit.
- The company's responsibilities extend from the earliest stages of exploration to well beyond the life of the mine itself.
- Mining operations have significant environmental impact that may affect resources used by local and indigenous communities.
- Mining companies employ local people but also attract workers from elsewhere. This can create tensions among stakeholders, potentially jeopardising the company's social licence to operate and the stability needed to operate successfully.

AngloAmerican has issued annual SRs since 2000 following the guidelines of the GRI, described earlier. The 2012 Report acknowledges that mining can have significant environmental impacts: it uses large amounts of energy, generates green-house gases, needs a great deal of water and creates waste that requires containment and treatment. The SR identifies six key sustainability areas – water, health, safety, community development, climate change and operational excellence – that represent risks to be managed but can also become opportunities to create value for society. The report lists the stakeholders, the mechanisms in place to engage with them, the concerns they express and the actions the company has taken or will take to respond to those concerns. It compares its performance in the current year with that in the previous year, demonstrating (in the 2012 report) significant progress in water and energy conservation, community development programs and investment in local housing and education.

Independent assurance of key performance elements of the Report is provided by PricewaterhouseCoopers and is highly ranked by the Dow Jones Sustainability Index.

[9]http://www.angloamerican.com/~/media/Files/A/Anglo-American-Plc/reports/AA-SDR-2012.pdf.

CASE STUDY 2: A MATERIAL USER

Jaguar Land Rover[10] is a maker of luxury cars. Cars are responsible for about 20% of all the carbon released into the atmosphere[11]. The auto industry is under pressure from National governments to reduce tailpipe emissions and a sector of the public, too, perceives the emissions from large vehicles such as those made by Jaguar Land Rover to be irresponsible. The average life of a car is only 8 years[12], generating further pressure to improve their recyclability. The pressures take the form of

- Financial penalties if increasingly stringent fleet economy and emissions standards are not met.
- Legislation requiring that 85% of the vehicle be recycled at end of life.
- Concerns for potential materials supply-chain and water constraints.

Jaguar Land Rover first started issuing SRs in 2009. Not surprisingly, the 2011/2012 report emphasizes the steps the company has taken to improve its environmental performance. The report documents the weight reduction of their vehicles over past years made possible by the extensive use of aluminum and plastic. It details the reduced carbon footprint of manufacture by increased recycle content of the aluminum and the use of bio-reinforced plastic for panels and trim, and of "certified low-carbon leather" for seats. The company plans further weight reduction and improved eco-performance, aiming for 75% recycle content in the structural aluminum components and the use of CFRP to achieve further weight saving.

The CEO Statement of the Jaguar Land Rover SR emphasises the commitment of the company to add value to the communities in which it operates through training programs. Its global CSR programme highlights a carbon-offset programme that has compensated for 5 million tonnes of CO_2 to date via wind, hydro and other renewable energy plant in Africa, Asia, Europe and South America. The company invests in projects that have additional social and economic benefits; these, they claim, have helped improve the lives of over 1.2 million people.

They donate over £1m per year in cash and kind to charities. The company was awarded the highest (Platinum) level in the Community CR Index.

[10]http://www.landrover.com/imagery/global/downloads/sustainability-report/sustainability-report-2011-2012.
[11]www.epa.gov/climatechange/ghgemissions/sources.html.
[12]http://www.nbcnews.com/id/12040753/#.UpTK3mfwjK0.

6.4 SUMMARY AND CONCLUSIONS

Over much of the last century material prices, in real terms, were static or falling. There was relatively little control over the way materials were used or what happened to them at the end of product life. Corporate priorities focussed on short-term profitability and financial returns to shareholders.

Today the picture looks rather different. As we saw in Chapters 1 and 5, the increasing complexity of products creates a dependence on a larger number of elements, some comparatively rare. These are sourced globally and used to make products that are traded on a global scale. Manufacturing nations increasingly compete for preferential or exclusive rights to mineral resources worldwide in order to safeguard their industrial capacity. New and expanding legislation controls many aspects of manufacturer responsibility, product design, material usage and material disposal. The public, stakeholders and government increasingly judge corporate success not just in financial terms but in terms of stewardship of the environment and welfare of its workforce and that of the local economy of the communities in which it operates. Corporations respond by issuing SRs detailing their attention to CSR. Sustainability reporting serves several purposes. Externally, it increases organizational transparency and informs stakeholders of company sustainability strategy. Internally, it raises awareness of the need for change and the steps that are planned to achieve it. A robust, defensible sustainability policy has become a key element of business planning.

6.5 EXERCISES

E6.1 An influential investor in a small company making household products insists that the company issue a SR, following the GRI guidelines. Briefly, what is the GRI?

E6.2 The CEO of a small company making specialized copying equipment becomes aware that competitors now issue Environmental Product Declarations (EPDs) for their products. His Board is concerned that failure to follow suit may lead to loss of sales. What is an EPD?

E6.3 The web site CSRwire (http://www.csrwire.com/reports) gives access to SRs from a large number of companies. Select one of the companies that either provides or uses materials, survey the report and write a half-page summary like that of the examples in the text of what you find. Use the criteria listed in Section 6.3 for guidance.

E6.4 Emerson is a manufacturing company that uses materials on a large scale. Track down the Emerson SR for the previous year (that for the current year will not yet be complete), survey its content and write a half-page précis of the points that strike you as significant in Emerson's operations.

E6.5 Look up the SR from Apple. Summarize it. Write a list of the stakeholders mentioned in it. Identify any claims made to do with materials choice. Which countries are mentioned in the report? Are they countries you think will be good places to work with over the next 3 years? What timescales are mentioned in the report? Is Apple thinking long term?

Further Reading

AngloAmerican Mining Company Sustainability Report, 2012. http://www.angloamerican.com/~/media/Files/A/Anglo-American-Plc/reports/AA-SDR-2012.pdf.

Community's Corporate Responsibility (CR) Index, 2014. http://www.bitc.org.uk/our-services/benchmarking/cr-index (The index ranks companies on their environmental stewardship, their social contributions and their business practice.)

CSRwire, 2014. http://www.csrwire.com/reports (The site gives access to Sustainability Reports from a large number of companies.)

Dow Jones Sustainability World Index, 2014. http://djindexes.com/sustainability/?go=literature and http://www.sustainability-indices.com/ (The index is based on consistent improvement in a company's strategies for climate change, energy consumption, human resources development, knowledge management, stakeholder relations and corporate governance as reported in their SR.)

Ethical Corporation, 2013. Full Product Transparency Is the Future of Reporting. http://www.ethicalcorp.com/communications-reporting/full-product-transparency-future-reporting (Guidance on preparing Sustainability Reports.)

FTSE4Good Index, 2014. http://www.ftse.co.uk/Indices/FTSE4Good_Index_Series/index.jsp (The index claims to be an objective measure of the performance of companies in meeting globally recognised corporate responsibility standards.)

Granta Design, 2013. Enabling Product Design and Development in the Context of Environmental Regulations and Objectives, a White Paper. http://www.grantadesign.com/emit/ (Granta Design is a company specializing in material information management; its EMIT Consortium advises on environmental strategies and risk management.)

Jaguar Land Rover Sustainability Report, 2012. http://www.landrover.com/imagery/global/downloads/sustainability-report/sustainability-report-2011-2012.

CHAPTER 7

Introduction to Case Studies

Chapter Outline

7.1 Introduction and Synopsis 111
7.2 The Structure of the Case
   Studies 112
7.3 Articulations of Sustainable
   Development That Went
   Wrong 113

7.4 Summary and Conclusions 115
7.5 Exercises 116
Further Reading 116

7.1 INTRODUCTION AND SYNOPSIS

The six chapters that follow this one are case studies. Each investigates an articulation of sustainable development using the 5-step method developed in Chapters 3 and 4. The topics are as follows:

- Biopolymers to replace oil-based plastics
- Wind farms as a source of renewable energy
- Electric cars as the future for clean personal transport

Materials and Sustainable Development. http://dx.doi.org/10.1016/B978-0-08-100176-9.00007-4
Copyright © 2016 Elsevier Ltd. All rights reserved.

- Lighting – incandescent, fluorescent, LED
- Solar PV for low-carbon power
- Bamboo as a sustainable building material

They are designed for project-based teaching.

7.2   THE STRUCTURE OF THE CASE STUDIES

The case studies follow a standard pattern. Each starts with an Introduction and Background information, providing context and basic, relevant, facts. This leads into the 5-step analysis. Each step is identified by an icon, shown here on the right.

Step 1 is a statement of the *Prime Objective, size scale and time scale*. Sometimes these are explicitly stated in the description of the project. When they are not, we need common-sense estimates.

Step 2 is the identification of *Stakeholders and their concerns*, ideally through interviews and questionnaires. When this is not possible, the National Press plus a web-based search of news reports, Government and NGO position-papers, letters to editors and magazine surveys reveal a great deal about them. The interest and influence of each stakeholder is mapped onto the stakeholder diagram.

Step 3 is *Fact-finding*, using literature and internet searches, to research both the factual background of the articulation and the validity of the concerns expressed by the stakeholders. Facts should, if possible, be checked against a second, independent source. All the facts assembled in the Fact-finding step of the Case Studies are available from open sources but they are scattered. It helps to re-visit the checklists of Chapter 4 at the start of the Fact-finding step. The Appendix of Chapter 17 is an assembly of facts relating to materials, energy and national characteristics; a paragraph highlighted in blue in each case study indicates how this information is found and used. Fact-finding is made easier by using the CES EduPack Sustainability database described in Chapter 15, but this is a convenience not a necessity.

Step 4 *Synthesis* is an examination of how the facts influence Human, Natural and Manufactured Capital. Some facts have a positive impact on one capital but a negative one on another. Articulations designed to reduce harmful emissions, for example, have a positive influence on Natural Capital but may incur expense and

thus create a burden on Financial Capital. Mapping the findings of the Fact-finding step onto the Synthesis matrix shown in Table 4.5 helps to assemble a big picture. In the Case Studies, a "fact" that has a generally positive influence is given a (+) and is colored green, one that has a generally negative influence is given a (–) and is colored red. The same fact can appear in more than one cell of the matrix. As an example, the use of corn as a feedstock for biopolymer production might carry a (+) for Natural Capital but it might carry a (–) for Financial Capital because it drives up corn prices.

Step 5, the final step, is that of informed *Reflection*, assessing the picture that has emerged and considering priority changes. It is valuable to do this with two timescales in mind. Many articulations appear, in the short term, to be impractical or misguided because the infrastructure essential for their success is not in place and cannot be created in the short time frame visualized when the articulation was first formulated. Decarbonizing the national electricity grid to allow carbon-free charging of electric cars is clearly something that cannot be done in a 10-year time frame. In the longer (40-year) term, however, this might be possible. Thus many present-day articulations of sustainable development are, as it were, first steps on a longer journey. Does the long-term gain justify the short-term pain? Step 5 is also the moment to ask: Could the vision behind the original articulation be achieved in a less-painful way than that first proposed?

The first three steps in this progression are objective and deterministic; the last two are subjective, and therefore open to debate. To reiterate a point made in Chapter 1: there is no "right" answer to assessing issues as complex as those of sustainable development. Instead there is a thoughtful, well-researched response that accepts the complexity and seeks to work with it to reach a balanced, fair and defensible conclusion.

7.3 ARTICULATIONS OF SUSTAINABLE DEVELOPMENT THAT WENT WRONG

Not all articulations of sustainable development succeed; indeed some fail in spectacular ways. We can learn something from these mistakes. Here are three examples.

THE GROUNDNUT SCHEME

The Groundnut Scheme was a British Government plan to cultivate peanuts ("ground nuts") in what was then Tanganyika, now Tanzania. The prime objective was to create a sustainable source of vegetable oil for Britain, at that time (1946) short of food. The Ministry of Food authorized £25 million to clear and cultivate 600 km^2 of land in the colony. Fact-finding (Step 3) for the proposed project was inadequate. Work started before soil samples were taken or water sources identified. Neither the climate and soil nor the lack of insects for pollination suited the peanuts. The project was abandoned after 5 years and an investment of £49 million, roughly £1000 million today, none of it recovered.

Wood, A. (1950). "The Groundnut Affair", London: Bodley Head.

TECHNOLOGY LOCK-IN

One way to deal with municipal waste is to burn it, using the recovered heat for district heating. One such project in Stockholm involved the construction of a plant fed by the local waste stream that was current at the time of construction. At the time this seemed attractive in two ways, that of waste disposal and low-cost heating for new apartment blocks, which were therefore built without any alternative heating system. As recycling in Sweden grew more efficient, less waste became available to feed the plant. The heating needs of the apartments remained the same. The plant is now fed in part by waste imported to Sweden from surrounding nations, 800,000 tonnes per year, requiring transport to the plant. The associated costs and environmental impacts make the scheme no longer attractive but the large initial investment and the absence of independent heating for the apartments locks it in. This is an example of a failure of Fact-finding, but one that is hard to anticipate: the missing fact (that the volume of municipal waste would decline because of more efficient recycling) was not known or foreseen at the time the plant was designed.

Mulder, K., Ferrer, D. and Van Lente, H. (2011) "What is sustainable technology" Greenleaf Publishing, UK.

CARBON-OFFSETTING GONE WRONG

Carbon off-setting is the buying of credits to compensate for the carbon emissions released by an industry or individual. Manufacture, power generation from fossil fuels, heating and some services (particularly transport) all have large carbon footprints. The idea is that money collected in off-sets be invested in activities elsewhere that absorb or sequester carbon or that replace the use of fossil fuels by energy sources that are carbon free: tree planting, solar, wind or wave power, for example. Much of the off-set investment is in less-developed areas of the world such as Africa, India and south-east Asia. By purchasing sufficient credits, the generator of CO_2 can claim to be "carbon neutral". There are controls in place that are designed to ensure that the money reaches its intended target.

Heather Rogers* tracks the path of money collected in this way. The money passes through more than one set of hands before it reaches its target. Her finding is that, often, little of it does; the hands absorb most of it as administrative costs or kick-backs. Here the problem is a human one: the idealism of the carbon off-set scheme was not matched by that of the chain that managed it.

*Rogers, Heather (2010) "Green gone wrong: How our economy is undermining the Environmental Revolution" Scribner Publishing.

7.4 SUMMARY AND CONCLUSIONS

This chapter sets up the 5-step structure of the case studies that follow. The first three steps are straightforward. The last two are not. You may not agree with the way the facts have been assigned to capitals in the Synthesis step of these case studies, or with the lines of reasoning that emerge from the final Reflection step. If you do not agree, it means you have a different view. Expressing these views clearly is one of the skills that this approach to assessing sustainable development can teach.

None of the articulations of sustainable development explored here is without objections – indeed, some appear to be impossible in the short term. It is important, in each case, to think also about the longer term and whether short-term sacrifice should be accepted to achieve long-term gain. It is appropriate here to end with a quotation:

"Nothing will ever be attempted if all possible objections must first be overcome"

Samuel Johnson (1709 – 1784)

7.5 EXERCISES

E7.1 Explore examples of articulations of sustainable development that went wrong, seeking to identify at which point the mistake was made. Was it because stakeholders' concerns were not met? Or because critical facts that could have been explored were in fact ignored (like the climate and pollination problem of the ground nut scheme)? Was it because the negative impact of the facts on one of the Capitals was overlooked? Was it because basic knowledge was lacking, so that the consequences of the articulation, when implemented, could not have been foreseen (like ozone-depleting effect of CFCs)? You will find examples of articulations that went wrong in the two books listed under Further Reading.

E7.2 Research the history of CFCs as a heat-transfer medium for refrigeration and air-conditioning. At what point in the assessment of CFCs was the mistake made that led to their introduction? (They are now banned.)

Further Reading

Rogers Heather: *Green Gone Wrong: How Our Economy Is Undermining the Environmental Revolution*, Scribner, 2010. ISBN13: 9781416572220; ISBN10: 1416572228 *(A readable, first-hand account of the many ways in which well-intentioned green projects can go wrong.).*

Mulder K, Ferrer D, Van Lente H: *What Is Sustainable Technology*, Sheffield, UK, 2011, Greenleaf Publishing, ISBN: 978-1-906093-50-1 *(A set of invited essays on aspects of sustainability, with a perceptive introduction and summing up.).*

CHAPTER 8

Scaling Up Biopolymer Production

Chapter Outline

8.1 Introduction and Background Information 118

8.2 Prime Objective and Scale 120

8.3 Stakeholders and Their Concerns 120

8.4 Fact-Finding 122

8.5 Synthesis with the Three Capitals 127

8.6 Reflection on Alternatives 130

8.7 Related Projects 132

Further Reading 133

Materials and Sustainable Development. http://dx.doi.org/10.1016/B978-0-08-100176-9.00008-6
Copyright © 2016 Elsevier Ltd. All rights reserved.

8.1 INTRODUCTION AND BACKGROUND INFORMATION

Commodity plastics are made from fossil hydrocarbons, particularly oil. This is a concern for the following four reasons:

- oil is a nonrenewable resource and a precious one;
- using fossil hydrocarbons ultimately releases the carbon they contain to the atmosphere;
- dependence on oil exposes the industry to risk of cost volatility and supply constraints, and
- most oil-based polymers degrade only very slowly creating a long-term problem of "polymer pollution" (Figure 8.1).

Biopolymers, by contrast, are plastics made from biomass. Hydrocarbons derived from renewable sources such as corn, soya, cellulose and polysaccharides such as sugarcane can be polymerized to make bio-PE, bio-PET and other less familiar plastics such as PLA and PHA. Today they are used for packaging (NatureWorks PLA), disposable cutlery and containers (Cereplast's PLA blend, Novamont's Mater-Bi starch resin), dental care items and medical items (Cereplast) and agricultural turf stakes (Telles PHA).

Biopolymers are promoted as sustainable substitutes for plastics derived from oil. If they are to do this they must *really* be more sustainable than oil, must perform as well as the materials they displace, be affordable and producible on a scale that compares with that of commodity plastics.

FIGURE 8.1
Biodegradable biopolymer products.

The Bio-based Society[1] makes the following claim: *"Biobased polymers will increase in capacity to 12 million tons/year by 2020. That will equal 3% of total polymer production."* That is a statement of an anticipated sustainable development. What is the background? Who are the interested parties? Is it a good idea to increase biopolymer production? How do the carbon footprint and price of biopolymers compare with those made from oil? In short, is it a sustainable development?

8.1.1   Background Information

- Commercial biopolymers are listed on the left of Table 8.1. Competing oil-based polymers are listed on the right. Not all biopolymers are biodegradable. Those that are, are starred (*).
- The global production of plastics in 2011 was 280 million tonnes, growing at 4% per year; thus the expected production in 2020 is about 400 million tonnes. Almost all are derived from oil, consuming about 5% of global oil production.
- The global production capacity for bioplastic was a little over 1 million tonnes in 2011,[2] or 0.25% of all polymer production. The strongest growth is expected in the bio-based, non-biodegradable bioplastics group, especially the so-called

Table 8.1 Biopolymers and Commodity Oil-Based Polymers	
• **Biopolymers** •	• **Oil-Based Polymers** •
Bio-PP (made from ethanol)	PP (Polypropylene)
Bio-PE (made from ethanol)	PE (Polyethylene)
PLA (Polylactic acid)*	Polyvinylchloride
PHA (Polyhydroxyalkanoate)*	Polyurethane
PTT (Polytrimethylene terephthalate)	PET (Polyethylene terephthalate)
CA (Cellulose acetate)	PS (Polystyrene)
PA11 Nylon 11	PA6 Nylon 6
TPS (Thermoplastic starch)*	PCL (Polycaprolactone)*

*PLA, PHA, TPS, PCL and blends of these with PE, PP and PET are biodegradable.

[1]http://www.biobased-society.eu/2013/03/biobased-polymers-will-make-a-breakthrough-within-ten-years/.
[2]http://en.european-bioplastics.org/market/market-development/production-capacity/.

'drop-in' solutions, i.e. bio-based versions of bulk plastics like PE and PET, synthesised from bioalcohol.

- The predominant feedstock for biopolymer production is corn. Growing 1 kg of corn per year requires roughly 1.25 m² of fertile land. It takes 3 kg of corn to make 1 kg of ethanol. It takes 2 kg of ethanol to make 1 kg of polyethylene. Thus it takes 7.5 m² of fertile land area to grow 6 kg of corn to make 1 kg of biopolyethylene per year. Globally, the potentially productive land area[3] is 31 million km²; that of the United States is 1.6 million km²; that of Europe is 1.1 million km².

Resources in the Appendix. Table A7 gives price data for polymers, including biopolymers. Table A18 lists embodied energy, carbon footprint and water usage of the same polymers.

8.2   PRIME OBJECTIVE AND SCALE

Prime objectives, size and timescale are defined in the project statement: a global production of 12 million tonnes of biopolymers per year by 2020.

8.3   STAKEHOLDERS AND THEIR CONCERNS

Here are five quotes from people and organizations that have an interest in biopolymers.

- "Bioplastics make good business sense. All other things being equal, independent market research has shown a majority of customers would patronize a business that is "green" over one that isn't.[4]" (Biomass Packaging, 2013)
- "'Sustainable' bio-plastic can damage the environment. Corn-based material emits climate change gas in landfill and adds to food crisis[5]" (The Guardian, 26 April, 2008)
- "Finding alternative sources for materials is becoming imperative as petroleum prices continue to rise[6]" (John Viera, Ford's

---

[3]https://en.wikipedia.org/wiki/Arable_land.
[4]www.biomasspackaging.com/education/bioplastics/.
[5]www.guardian.co.uk/environment/2008/apr/26/waste.pollution.
[6]www.thingsaregood.com/tag/bioplastics/#sthh.SJYD8jYY.dpuf.

global director of sustainability and vehicle environmental matters, May, 2012)

- "Worldwide demand for bio-plastics will grow fast over the next decade….(driven by)…government regulation and legislation to promote sustainability and bio-degradability[7]" (NanoMarkets report, January 2013).
- "The EuPC is concerned by statements that bio-based plastic shopping bags are more sustainable than oil-based alternatives. Bioplastics are not a solution to marine litter.[8]" The EuPC is the EU-level trade association representing European plastics converters.

Even these five quotes suggest the main interested parties (Figure 8.2). A little further research gives this list.

- Government, who see biofuels and biopolymers as ways of reducing dependence on oil and reducing carbon emissions[9] and polymer pollution.
- Polymer producers, who see biopolymers as a growing and potentially profitable market.
- Farmers, who can profit by diverting farm production to crops for biofuels and biopolymers.[10]

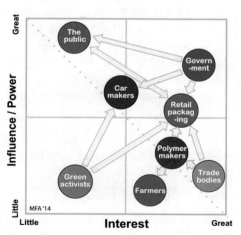

FIGURE 8.2
Stakeholders and Influence.

[7]http://nanomarkets.net/blog/article/regulatory_drivers_for_the_bio_plastics_market.
[8]www.europeanplasticsnews.com/subscriber/newscat2.html?cat=1&channel=620&id=3594.
[9]https://bioenergy.ornl.gov/main.aspx.
[10]Ibid.

- The food packaging industry, supermarkets and other retailers. A survey[11] of their concerns listed "Sustainability", "Economy" and "Product value promise (aesthetics and perception)" as the three most important.
- Consumer associations, alarmed at overstated claims of "greenness", who want truth in advertising.
- Car makers see biopolymers as a visible demonstration of their efforts to reduce the environmental impact of cars but express concern that there may be a weight and cost penalty in using them.[12]
- Green campaigners[13] who fear the increased use of biopolymers will contribute to the global food crisis by diverting fertile land from food to biopolymer-feedstock production.
- Concerned public who see a need to reduce oil-dependence for geopolitical and economic reasons.
- Consumers, who must be persuaded to select biopolymer products.

SUMMARY OF THE SIGNIFICANT CONCERNS

- Are the biopolymer properties and processability as good as those of oil-based plastics?
- Can biopolymers reduce dependence on fossil hydrocarbons?
- Can they reduce carbon emission associated with polymer production?
- Will biopolymer production conflict with the production of food and feedstock?
- Will biopolymers help reduce polymer pollution?

These concerns and the claims made by the Bio-based Society flag the targets for research in the fact-finding stage. That comes next.

8.4 FACT-FINDING

What information is needed to support or refute the claims made for biopolymers and the concerns expressed about them? What additional facts do we need for a rational discussion of the Prime

[11]Kingsland, C. (2010) "PLA: A critical analysis" Packaging Digest www.iopp.org/files/public/KingslandCaseyMohawk.pdf.
[12]www.europeanplasticsnews.com/subscriber/headlines2.html?id=1759.
[13]http://www.guardian.co.uk/environment/2008/apr/26/waste.pollution.

Objective? Figure 8.3 summarizes the issues needing research. No judgements yet; just facts.

1. **Material attributes.** Figure 8.4 shows two mechanical properties, Young's modulus and Tensile strength, each plotted against density. (The data shown here are for unfilled, moulding-grade materials.) Biopolymers, generally, have stiffness and strengths that are comparable with those of PE or PP, but they are heavier, which reduces the values of stiffness and strength per unit weight. Biopolymers have low heat resistance and poor impact resistance, both of which can be improved with plasticizers and by blending with oil-based plastics, but this inhibits recycling.

   Biopolymers are more difficult to mold than oil-based polymers because of the narrow window between the processing temperature and the decomposition point.[14] It is practical to make film, rigid packaging and small (up to 250 grams) injection molded parts, but larger parts are problematic.

2. **Energy.** Figure 8.5 compares the embodied energy[15] per unit volume of biopolymers and oil-based plastics, using today's

FIGURE 8.3
A Fact-finding summary. Center: a PLA plate set, courtesy SDA.

[14]http://www.ptonline.com/articles/injection-molding-bio-polymers-how-to-process-renewable-resins.
[15]The "embodied energy" of an oil-polymer includes both the energy required to synthesize it from oil and the energy content of the oil itself (roughly 44 MJ/kg).

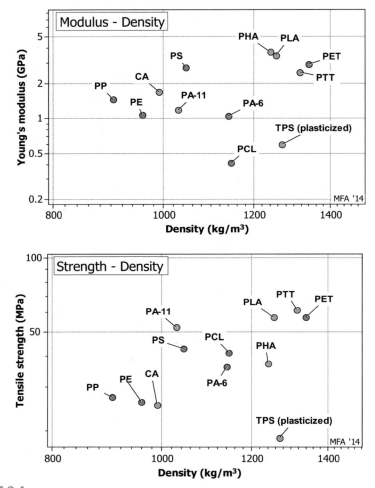

FIGURE 8.4

Young's modulus, E, tensile strength, and density ρ of commodity oil-base polymers (red) and biopolymers (green), with contours of E/ρ and σt/ρ.

feedstock and processing methods. There is very little difference between them – only starch-based polymers (TPS) have significantly lower embodied energy. At first sight, it seems surprising that a polymer based on natural materials is nearly as energy-intensive as one made from oil. It is because the fermentation or processing needed to make bioresins requires heat and thus carries an energy and carbon burden, and the subsequent polymerization step for bio- and oil-based polymers makes almost identical contributions to both.

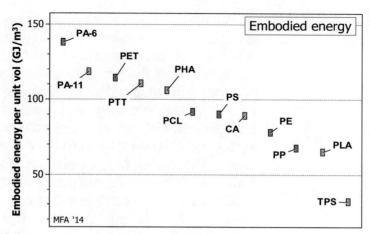

FIGURE 8.5
The embodied energy of oil-based polymers (red) and biopolymers (green).

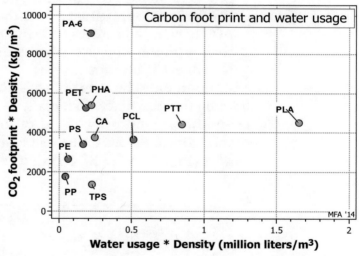

FIGURE 8.6
The carbon footprint and water usage of oil-based polymers (red) and biopolymers (green).

3. **The Environmental properties.** Biopolymers are widely perceived to have a better eco-character than oil-derived polymers, but evidence does not entirely bear this out. Figure 8.6 shows the carbon footprint and water usage of bio- and oil-based polymers. Today's biopolymers compare poorly with conventional oil-based alternatives. The water demand of biopolymers is high because of the needs of the plant or animal feedstock from which they are derived.

What about polymer pollution? Some biopolymers biodegrade in the true sense of returning their constituent to the biosphere in the form from which they were first drawn. But not all. More than half the present day production are "drop-in" biopolymers (those identical in chemistry with oil-based plastics) such as bio-PE, and these, of course, do not biodegrade.

4. **Legislation.** The largest single use of bioplastics is in packaging and throw-away food items. Most nations have legislation that controls aspects of the packaging industry. Much of it seeks to encourage recycling and the use of biodegradable or compostable materials, providing incentives to use the bioplastics that meet these criteria. Here are two examples.

   - EU Packaging Directive 94/62/EC, which urges the use of biodegradable or compostable packaging.
   - EU Registration, Evaluation, Authorization and Restriction of Chemical Substances Directive EC. 1907/2006 (REACH) that prohibits the use of certain plasticizers and flame retardants in plastics.

   It is important to be aware of legislation like these when considering the use of new materials.

5. **Economics.** A price for crude oil of $100 per barrel corresponds to a cost per unit weight of $ 0.72/kg. Commodity plastics (PP, PE, PVC, PET, PS) all have prices between $1.5 and

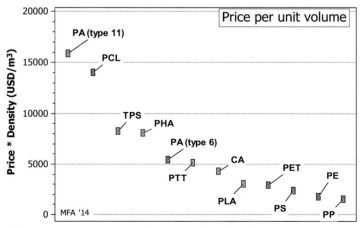

FIGURE 8.7
The price per cubic meter of oil-base polymers (red), and biopolymers (green).

$2.8 per kg, only about 2 to 4 times that of the oil on which they are based, making them vulnerable to rising oil prices.

Biopolymer, today, cost more than commodity oil-based polymers like polyethylene and polypropylene. Figure 8.7 plots the price per unit volume (in a straight substitution of a biopolymer for a conventional polymer, it is the price per unit volume that is significant). PLA is about 25% more expensive per unit volume than PP, PE or PS. All the other biopolymers are more expensive than this.

6. **Society.** One of the stakeholder concerns was that of "Product value promise (aesthetics and perception)". A sustainable development must gain public acceptance. Market appeal is partly a question of industrial design. Cellulose acetate and PLA are available in grades with high clarity and gloss that make for attractive food packaging. The association with environmental stewardship by severing the link with petrochemicals and offering biodegradability has emotional appeal. However, some biopolymers have a smell that makes them less appealing for durable products such as car interiors.

SUMMARY OF SIGNIFICANT FACTS

- Biopolymer properties are comparable with those of oil-based plastics but they are more difficult to process.
- Biopolymer production on a significant scale from cultivated crops requires a large area of fertile land.
- Biopolymers are, at this point, more expensive than PE, PP, PS or PET.
- Biopolymers have a lower carbon footprint than oil-based polymers, but not by much.
- Some biopolymers are biodegradable, but not all. The ability of biopolymers to reduce polymer pollution has sometimes been overstated.

8.5 SYNTHESIS WITH THE THREE CAPITALS

What impact do the facts have on the three capitals? Some affect more than one capital; for simplicity, we assign the influence to the one most affected. As explained in Chapter 3, there is no single answer to this question; the weight given to the facts depends on

Table 8.2	Synthesis: influence of the facts on the three capitals		
	Human and social capital – People *Health? Wellbeing? Convenience? Culture? Tradition? Associations? Perceptions? Equality? Morality?*	**Natural capital – Planet** *Can prime objective be met? Are stakeholder concerns addressed? Are there unwanted consequences?*	**Manufactured capital – Prosperity** *Cost – Benefit? (Cost facts vs. Eco facts) Legitimacy? Conformity with law?*
Materials	(+) Possibility of establishing plastics-based manufacturing in countries without oil but with land and water. (+) Emotional appeal of bio-polymers to some consumers becomes a design feature	(+) A reduction in demand for oil possible (-) The reduction is small (0.15% for 3% replacement; 1.5% for 30% replacement) (-) Large increase in water demand (15% for 3.% replacement, 150% for 30% replacement)	(-) Properties and processability less good than oil-based plastics (+) Non-dependence on oil makes cost less sensitive to changing oil prices (-) Dependence on corn, soya or sugar cane makes cost of bio-polymers sensitive to changing commodity prices
Energy		(+) Reduction in oil consumption small but real	(-) Bio-polymers are more expensive than oil-based polymers at present
Environment		(+) Some bio-polymers are biodegradable (-) Carbon footprint not significantly less (-) Increased demand for water (-) Requires considerable area of fertile land	
Legislation	(+) Reducing public waste increases quality of life	(+) Recycling and composting legislation reduces waste stream	
Economics	(+) The winners: farmers, with broader market (-) The losers: consumers – higher food prices		(-) Bio-polymers at present more expensive, harder to process, than oil-based polymers (+) A growth industry, despite cost
Society	(+) Creates "green" jobs (+) Some satisfaction in using renewable materials (-) Diverts fertile land from food production		(+) Potential to stimulate local industries such as 3-D printing in oil-poor countries
Synthesis (the most telling facts)	(?) Mistaken appraisal and emotional response to bio-polymers sustains the market at present.	(+) Has potential to reduce waste stream (-) No real saving of oil; no significant reduction in carbon, considerable increase in water demand.	(-) Bio-polymers not, at present, cost-effective – but they are nonetheless accepted and profitable. (+) Increased scale of production, improved properties, could reduce cost.

values, culture, beliefs and ethics. Here we present one view. The main points are summarized in Table 8.2; positive influence in green, negative influence in red. The text and the table that follow give details.

8.5.1   Natural Capital

- *Reduced dependence on fossil hydrocarbons?* Polymer production uses about 4% of the global production of fossil hydrocarbons. Replacing 3% of this by biopolymers reduces demand by 0.12%. Even the ideal of 30% replacement reduces the demand by just 1.2%. So – yes, but not much.
- *Reduced carbon emission?* The carbon footprint of biopolymers is marginally different from that of oil-based polymers. There is no evidence, at present, that biopolymers make a significant difference.
- *Conflict with the production of food and animal feedstock?* The land area required to synthesize conventional polymers is negligible; that for biopolymers is large, up to $7.5 \, m^2/kg$ per year. At that level, the production of 1 million tonnes of biopolymer per year uses $7,500 \, km^2$ of fertile land; 12 million tonnes needs $90,000 \, km^2$. This is about twice the area of either the Netherlands or the State of Massachusetts. The potentially usable arable land area of the earth is about 31 million km^2[16] so this is about 0.3% of global productive land area.
- *Will biopolymers help reduce polymer pollution?* Some biopolymers can biodegrade but landfill sites are anaerobic, inhibiting the breakdown processes. Thus biodegradability may reduce polymer pollution at sea or in the countryside, but it will not reduce the volume of plastic in landfill.

8.5.2   Manufactured and Financial Capital

- Are biopolymers financially viable? Basing them on natural feedstock decouples their price from that of oil, but links it to that of commodities such as corn, the price of which also fluctuates with growing conditions and with market forces. Despite their higher price at present, the bioplastics market is

---

[16]https://en.wikipedia.org/wiki/Arable_land.

growing at 8–10% per year and is predicted by some studies to take up to 25% of the polymer market by 2020.[17]

■ Is the processability of biopolymers as good as those that are oil-based? Some biopolymers – those synthesised from bio-alcohols (bio-PE, for example) – are identical to their oil-based equivalents. Others are more difficult to process.

8.5.3   Human and Social Capital

■ *Personal satisfaction.* The interaction with Human Capital is psychological rather than real. Biopolymers, to some, offer a significant emotional satisfaction in a dependence on nature rather than fossil hydrocarbons.

■ *New industries.*[18] The availability of new materials (the biopolymers) could stimulate new industries (such as 3-D printing of polymer products) in land-rich, oil-poor countries.

■ *Are the public accurately informed?* There are indications[19] that the market for biopolymers is sustained by overstated claims for their sustainability. Building social capital depends on trust made possible by responsible public information.

8.6   REFLECTION ON ALTERNATIVES
8.6.1   Short Term

Can the Prime Objective be met? The current (2013) production of biopolymers is about 1 m tonne/year. Expanding it to the target of 12 m tonnes/year by 2020 means a growth rate of about 25% per year. This is twice the current growth rate of biopolymer production (10–12% per year) but is not impossible.

Do biopolymers cut carbon emissions or reduce plastic waste? Not significantly. Do they reduce dependence on oil? Again, not much. About 5% of oil production is used to make polymers. Replacing 3% this by biopolymers reduces oil demand by 0.15%. Biopolymer production is, of course, in its infancy; as production is scaled up and processing is refined it is expected that biopolymers will begin to show an environmental gain.

---

[17]www.hkc22.com/bioplastics.html.
[18]http://www.3ders.org/articles/20120208-developing-sustainable-bioplastics-for-3d-printers.html.
[19]www.europeanplasticsnews.com/subscriber/newscat2.html?cat=1&channel=620&id=3594.

The economics: the present "favourite child" status of biopolymers allows an expanding market even though they cost more than oil-based plastics. Research, development and scale-up is expected to make them cheaper and the rising cost of petrochemicals may disadvantage oil-based plastics. The advantage may be lost if the competition for land and water to grow the feedstock for biopolymers drives up commodity prices. And there is the uncertainty about supply – a poor growing season can halve the yield per hectare.

The evidence suggests that the goal of increasing global production to 12 million tonnes of biopolymers per year is achievable. But that production is buoyed up, for now, by an emotional response from the public and from government to our disquieting dependence on fossil hydrocarbons and the seeming purity of a dependence, instead, on nature. Making this a sustainable development in the short term depends on investment and trust.

8.6.2   Longer Term

Three issues will bite if biopolymers are to replace oil-based plastics on a significant (30% or more) scale; their price, their more limited range of properties and processability and their competition for resources with food production. In a world with many people without adequate food and water, diverting these to polymer production does not seem right. Could this not be overcome by seeking an alternative feedstock? There are at least two potential sources that do not compete with food: agricultural waste and marine algae.

Agriculture is the world's greatest industry and it produces waste on a corresponding scale. This waste could, potentially, eliminate the need for land and water for biopolymers. The global production of straw, for example, is comparable with that of grain. The challenge is to find ways to make such waste into a feedstock for the polymer industry.[20]

Marine algae[21] require only seawater, sunlight, carbon dioxide and nutrients to flourish, leaving freshwater and land for food

---

[20]One of many sources: http://www.worldchanging.com/archives/004756.html.
[21]Here is one (of many) sources. http://www.sciencedaily.com/releases/2012/10/121015084649.htm.

production. Algae multiply fast, producing up to 15 times more organic matter per unit area than land biomass. Fermenting either one to make ethanol and other alcohols could allow synthesis of bio-polyethylene, bio-PVC and bio-EVA.

Success in reaching the destination of this articulation appears to lie in investment in research and development in these two branches of biotechnology.

8.7   RELATED PROJECTS

These projects draw on some of the information assembled in the Biopolymer Production analysis, but the prime objective, the stakeholders, the operating requirements and expectations of users are all significantly different from those for biopolymers.

P8.1   **Biofuels.** The EU's Renewable Energy Directive (2009/28) sets an overall binding target of 20% for the share of EU energy needs to be sourced from renewables. As part of this effort, at least 10% of each Member State's transport fuel use must come from renewable biofuels by 2020.

Explore this articulation of a sustainable development. Biofuels are derived from the same feedstock as biopolymers, so many of the issues raised in this chapter apply to biofuels as well. The current consumption of energy for road transport in the EU zone is 384 million tonnes oil equivalent per year. The energy content of fuel for transport is 44 MJ/kg.

P8.2   **Biocomposites as nonstructural parts of vehicles.** A number of auto makers are trialing biocomposites for nonstructural components such as inner door panels and interior trim. The Prime Objective is to replace materials derived from nonrenewable resources (steel, oil-based plastics) with those that are renewable and (possibly) lighter. But does this articulation make sense? Are the properties of biocomposites as good as those they are supposed to replace? If they are not, sections will have to be thicker, so possibly heavier. Do biocomposites have a lower embodied energy than glass fiber reinforced sheet molding compound? Will consumers accept them (some smell a bit)? See what you can find out about them and then debate their contribution to the three capitals.

P8.3 **Could bioplastic be the answer to bottle woes?** (www.euronews.com 18 March 2014). Well, could they? Research the question, using the Case Study of this Chapter as a shortcut to some of the facts.

Further Reading

Wool RP, Sun XS: *Bio-Based Polymers and Composites*, Amsterdam, Netherlands, 2005, Elsevier, ISBN: 0-12-763952-7 (A text introducing the chemistry and production of bio-polymers.).

CHAPTER 9
Wind Farms

Chapter Outline

9.1 Introduction and
     Background 135
9.2 Prime Objective and Scale   138
9.3 Stakeholders and Their
     Concerns 138
9.4 Fact-Finding 140

9.5 Synthesis with the Three
     Capitals 145
9.6 Reflection on Alternatives   147
9.7 Related Projects 148
Further Reading   149

9.1 INTRODUCTION AND BACKGROUND[1]

The United Kingdom is legally committed to meeting 15% of its energy demand from renewable sources by 2020. The motives are to increase energy security and meet carbon reduction obligations (80% reduction in greenhouse gas emissions by 2050).[2] Current

[1]Image of wind farm courtesy of www.windjobsuk.com/wind-farm-jobs.cms.asp.
[2]www.gov.uk/government/policies/increasing-the-use-of-low-carbon-technologies.

Materials and Sustainable Development. http://dx.doi.org/10.1016/B978-0-08-100176-9.00009-8
Copyright © 2016 Elsevier Ltd. All rights reserved.

renewable energy generation (hydro, wind, solar, biofuel) is 12% of total demand, of which 6% is wind and 0.35% is solar. Growth of wind is 35% per year. If this growth is maintained for the next 7 years, the UK comes close to the 15% target.

Many nations have similar aims and strategies, encouraging the building of wind farms that feed electricity into the national grid. At the start of 2012 there were about 200,000 wind turbines worldwide, averaging 2 MW in power. The number, globally, is increasing at 25% per year, meaning that roughly 50,000 new turbines are installed each year. Is this a sustainable development?

Who are the stakeholders and what are their concerns? What materials, design, environmental, regulatory or social issues involved? To answer these questions we need facts. Armed with those, an opinion can be formed about the impact of wind farms on human, natural and manufactured capital. Given this information, a judgment be made of the contribution of wind farms to a more sustainable future.

9.1.1 Background Information

- Wind turbines produce energy only when the wind blows (Figure 9.1). The ratio of the actual average power output of the turbine divided by the nominal (rated) generating capacity is called the capacity factor. It is typically 20% for on-shore and 28% for off-shore turbines. Thus a 2-MW on-shore turbine will produce, on average, 0.4 MW.
- Most of the materials of a wind turbine are conventional: carbon steel, stainless steel, concrete, copper, aluminum and polymer matrix composites. One is exceptional. The generators of wind turbines use neodymium-boron rare-earth permanent magnets (Figure 9.2 and Table 9.1). Neodymium (also used in hybrid and electric vehicles) is classified in the US and Europe as a "critical" material. It is co-produced with other rare-earth metals, of which it forms 15% on average.
- A 2-MW wind turbine contains 25 kg of neodymium. Thus annual construction of 50,000 new turbines per year creates a demand for 1,250 tonnes of neodymium per year.

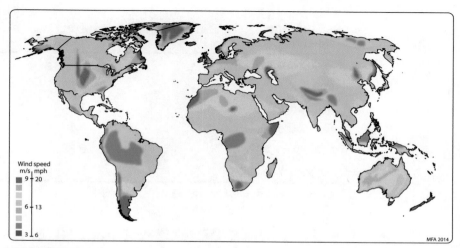

FIGURE 9.1
A global map of average wind speed.

Resources in the Appendix. Table A9 gives data for the energy intensity and carbon footprint of hydrocarbons. Table A12 lists the resource intensity (cost, area, material, energy, carbon) of power generating systems, including conventional gas and renewables, including wind. Table A13 reports the characteristics of energy storage systems.

FIGURE 9.2
The rotor of a permanent-magnet turbine.

Table 9.1 The Composition of Neodymium-Boron Magnets	
Nd-B Magnets	**Weight %**
*Neodymium (Nd)	30
Iron (Fe)	66
Boron (B)	1
Aluminum (Al)	0.3
*Niobium (Cb) (Nb)	0.7
*Dysprosium (Dy)	2

Starred (*) elements are on the critical list.

9.2   PRIME OBJECTIVE AND SCALE

Prime Objective and Scale are defined in the project-statement. They are to reduce global carbon emissions and increase energy self-sufficiency, with 2020 defining a time-frame. A target growth rate of 25% per year means building 50,000 new 2-MW (or equivalent, totaling 100 GW) units per year.

Is the objective realistic? To answer that, we need to know a number of things. What demands for materials will this building program create? Can they be met? By how much is the carbon footprint of electrical power from wind lower than that from fossil fuels? And will 100 GW of this lower carbon power make any significant reduction to global carbon emissions?

But first we must consider the stakeholders.

9.3   STAKEHOLDERS AND THEIR CONCERNS

A proper stakeholder assessment needs direct contact with those concerned. A start can be made by exploring the press and the Internet. Here are eight recent cuttings:

- *"Onshore wind: call for evidence"* (Title of UK Department of Energy and Climate Change consultative document, 6 June 2013).
- *"Government announces support for offshore wind energy industry"* (The Information Daily.com 1 August 2013).
- *"GE stimulates wind energy growth in the UK"* (Focus.com 5 November 2013).

- *"Marchers protest against wind farm plans"* (The Galloway Gazette, 28 November 2013).
- *"Planned onshore wind farms face uncertain future in Government shakeup that gives local communities a greater say over planning decisions"* (BusinessGreen.com, 6 June 2013).
- *"Centrica criticises policy as seabirds block Docking Shoal wind farm. Ministers block wind farm"* (The Telegraph, 6 July 2012).
- *"Strike a blow against wind-farm bullies"* (a columnist protests against the siting of wind farms in landscape he loves, The Times, 25 February 2013).
- *"Government and industry slam 'spurious' anti-wind farm headlines"* (The Times, 16 April 2012).

These headlines and the text that follows identify a number of stakeholders and their concerns (Figure 9.3). Among them are the following:

- *National and Local Government.* National Administrations have made commitments to reduce carbon emissions over a defined time period. They see wind farms as able to contribute. To encourage their construction they subsidize renewable energy production and impose taxes on carbon emission.
- *Energy providers.* Carbon taxes or carbon trading schemes and carbon penalties create financial incentives for energy providers to reduce the use of fossil fuels.

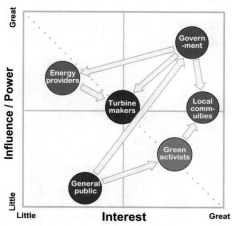

FIGURE 9.3
Stakeholders in planning a wind farm. The arrows suggest influence.

- *Wind turbine makers.* Turbine makers want assurance that government policy on renewable energy is consistent and transparent, that incentives will not suddenly be withdrawn and that the supply chain for essential materials is secure.
- *Local communities and the wider public.* The acoustic and visual intrusion of wind turbines and their power-distribution system is seen as unacceptable by some, as is the danger they pose for birds.

SUMMARY OF THE SIGNIFICANT CONCERNS

- The security of supply of the critical materials on which wind turbines depend.
- Wind farms do not reduce carbon emissions.
- Wind farms are uneconomic without subsidies; they drive up electricity prices.
- Wind farms are visually unacceptable.
- Wind farms are a danger to wild life.

9.4   FACT-FINDING

Step 3

What information is needed to analyse the claims made for wind farms and the concerns expressed about them? What additional facts do we need for a rational discussion of the Prime Objective – that of building 50,000 2-MW turbines per year? These questions are explored in the sections below. Figure 9.4 gives an overview.

Materials and supply chain.[3] Permanent magnets for electric turbines require high remanent induction with high coercive field.

---

[3]Ardente, F., Beccali, M., Cellura, M., Lo Brano, V., 2008. Energy performances and life cycle assessment of an Italian wind farm. Renewable Sustainable Energy Rev. 12 (1), 200–217. http://dx.doi.org/10.1016/j.rser.2006.05.013; Crawford, R.H., 2009. Life cycle energy and greenhouse emissions analysis of wind turbines and the effect of size on energy yield. Renewable Sustainable Energy Rev. University of Melbourne, Australia, 13 (9), 2653–2660; Danish wind industry association, 2003. What does a Wind Turbine Cost? http://guidedtour.windpower.org [accessed 08/10]; Martinez, E., Sanz, F., Pellegrini, S., Jimenez, E., Blanco, J., 2007. Life Cycle Assessment of a Multi-Megawatt Wind Turbine. University of La Rioja, Spain. www.assemblywales.org [accessed 08/10]; Vestas, 2008. V82-1,65 MW, Vestas Wind turbine brochure: Vindmølleindustrien. (1997, July). The Energy Balance of Modern Wind Turbines. Wind Power Note, 16. http://www.apere.org/manager/docnum/doc/doc1249_971216_wind.fiche37.pdf [accessed 08/11].

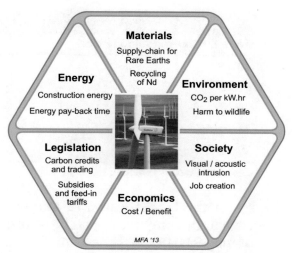

FIGURE 9.4
Fact-finding prompt for Wind farms.

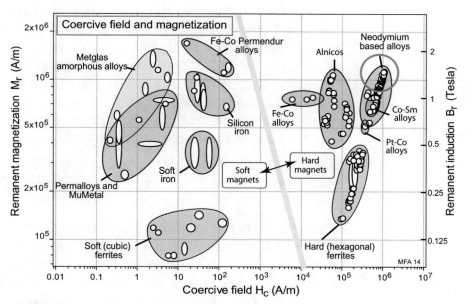

FIGURE 9.5
The remanent magnetization and coercive force of magnets. Nd-B magnets are ringed.

Figure 9.5 shows these two properties for magnetic materials.[4] Neodymium-based magnets (ringed in red at the upper right) have

---

[4]Ashby, M.F., Shercliff, H.R., Cebon, D., 2014. Materials: Engineering, Science, Processing and Design. Butterworth Heinemann, isbn:978-0-08-097773-7.

Table 9.2	Principal Sources for Rare-Earth Elements	
Rare Earth Producing Nation	**Tonnes/Year 2011**
China	130,000
India	3000
Brazil	550
Malaysia	30
World	133,580

Minerals.usgs.gov/minerals/pubs/commodity

by far the largest values of this pair of properties. If a substitute were to be sought, the next best choice would be the AlNiCo group of magnets, but all have a smaller remanent induction and a much smaller coercive field. Nd-B magnets are the current materials of choice for compact high-performance magnets.

Neodymium is co-produced with other rare-earth metals, of which it forms 15% on average. Table 9.2 lists the nations that produce rare-earths and the quantities they produce. The present global production is 133,600 tonnes per year, yielding 20,000 tonnes of Nd per year. Over 95% derives from a single nation, China. Nd is listed as a "critical" material because of its uniquely desirable magnetic properties for high-field permanent magnets, because its supply-chain is so narrow and because its price is volatile.

The current rate of building wind turbines described in the Introduction carries a requirement of 1,250 tonnes of neodymium per year. This is 6% of current global production.

Energy and energy pay-back time. Energy is used to make materials and to manufacture them into products. More energy is used transporting the products to where they will be used and assembling them to make a wind farm. Still more energy is used to connect the farm to the national grid. Numerous estimates[5] have been made of the energy required to build and commission a wind turbine – it is of general magnitude of 2×10^7 kJ per kW of nominal (rated) generating capacity.

How long will it take before a turbine has generated the energy that it took to make it? At a capacity factor of 0.2 it will take a time t:

---

[5]Ashby, M.F., 2013. Materials and the Environment. Butterworth Heinemann, and the CES EduPack 2014.

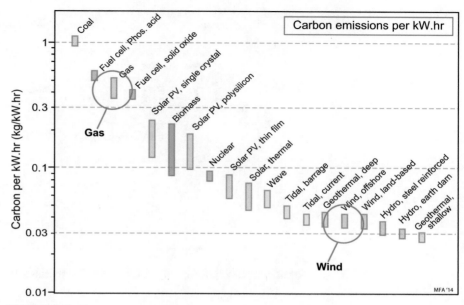

FIGURE 9.6
The carbon footprint of electrical power from coal, gas and low carbon sources

$$t \approx \frac{2 \times 10^7}{0.2} = 10^8 \text{ s} = 3 \text{ years}$$

The Environment. The Prime Objective of a wind farm is to generate electrical power with less carbon emissions than at present. It meets this objective only if the carbon emissions associated with its construction are more than offset by the low carbon emissions during life. Figure 9.6 compares the carbon emission per kilowatt hour of delivered power for alternative systems.[6] They are approximate, but sufficiently precise to establish that wind power has the ability to generate electrical power with significantly lower carbon emissions than gas- or coal-fired power stations when averaged over life. This, however, neglects power distribution: wind farms need windy places, often far from where the power will be used, and they may need energy storage systems to smooth intermittent generation.

Regulation. Much recent legislation across the world bears on reducing carbon emissions. They include carbon taxes, carbon

[6]MacKay, D.J.C., 2009. Sustainable Energy – Without the Hot Air. UIT Press, Cambridge, UK; Ashby, M.F., 2013. Materials and the Environment. Butterworth Heinemann, Oxford.

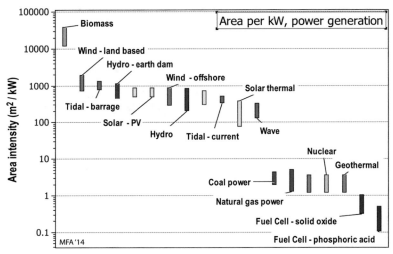

FIGURE 9.7
The area-intensity of power systems, using the Low-Carbon Power data-table.

trading and carbon off-setting. Making and installing wind farms is made financially attractive by "green" subsidies and feed-in tariffs but these have sometimes changed (usually down-graded) with little warning, making the market unpredictable.

Society. The manufacture and maintenance of wind farms creates jobs. If a proportion of the revenue generated by the farm is reinvested in the local community, it can build social capital as well.

Against this must be set the visual and acoustic intrusion caused by the turbines. Wind farms require a land-area per unit of generating power that is almost 1000 times greater than that of a gas-fired power station[7] (Figure 9.7) and while this land can still be used for agriculture the scale of the intrusion is considerable. To put this in perspective, if 10% of the electric power requirement of New York State (average 33 kWh per day, equivalent to 1.4 kW continuous per person, population 19.5 million) were to be met by wind power alone, the necessary wind farms would occupy 15% of the area of the entire State (area 131,255 km^2).

Economics. Are wind farms economic? Most of the commercial-scale turbines installed today (2014) are 2 MW (nominal) in size

---

[7]Ibid.

and cost between $3 million and $4 million[8] each. With a design life of 20 years, a load factor of 0.2 and allowing a sum equal to the cost of installation for life maintenance, ground rent and management, the cost of wind-farm electricity is $0.1 per kWh, somewhat more than that from a contemporary (2014) gas-fired power station. This, however, neglects the intermittency of wind power, which may create the need for energy storage. Grid-scale energy storage is expensive.

SUMMARY OF SIGNIFICANT FACTS

- Wind power can produce electrical power with significantly lower carbon emissions than gas- or coal-fired power stations.
- The construction energy of a wind farm is returned as electrical energy in 3–5 years.
- Compact, efficient turbines require neodymium, classed as critical, with significant supply-chain risk.
- The cost of energy from a wind farm is significantly higher than that from gas- or coal-fired stations. The cost rises further if grid-scale energy storage is needed to smooth the intermittent generation.
- Wind farms are intrusive to the communities in which they are sited. If wind farms are to contribute a significant fraction (say 10%) of energy needs, this intrusion becomes widespread.

9.5 SYNTHESIS WITH THE THREE CAPITALS

This is the moment to debate the relative importance of the information unearthed in the Fact-finding step, assessing its impact on the three capitals. It will, inevitably, require an element of personal judgement and advocacy. Here is one view, summarized in Table 9.3.

Natural Capital. The Prime Objective in building wind farms was to reduce green-house gas emissions. The studies cited above suggested that they can. The dependence on critical elements, particularly neodymium, for the turbines, might give concern but the placement of wind turbines is fixed and known, and large groups of them are managed by a single operator, making their recovery, reconditioning or recycling at end of life straightforward. Injury to bird life might be dismissed as trivial when domestic cats kill far more, but this is not a productive way to respond to stakeholder

[8]http://www.windustry.org/resources/how-much-do-wind-turbines-cost.

Table 9.3   Synthesis: Influence of the Facts on the Three Capitals for Wind Farms

	Human and social capital – People *Health? Wellbeing? Convenience? Culture? Tradition? Associations? Perceptions? Equality? Morality?*	Natural capital – Planet *Can prime objective be met? Are stakeholder concerns addressed? Are there unwanted consequences?*	Manufactured capital – Prosperity *Cost–Benefit? (Cost facts vs. Eco facts) Legitimacy? Conformity with law?*
Materials		(−) Large demand for critical elements, notably rare-earths	(−) Creates dependence on rare-earth producing nations (−) Requirement to create reuse and recycling infrastructure for rare-earth magnets
Energy	(+) Reduces dependence on imported energy by nations poor in fossil fuels	(+) Renewable power production	(+) Allows local autonomy of energy production (+) Creates employment
Environment	(−) Visual and acoustic interference diminishes quality of life	(+) Reduced carbon footprint of national electric power production (−) Harmful to bird life (−) On-shore wind farms require large area of land	
Legislation	(+) Helps meet the nation's commitments to reduce emissions		
Economics	(−) Need to subsidize wind farms adds to household energy bill		(−) Power from wind farms at present more expensive than power from gas or coal (−) Large capital investment that drains resources from other projects
Society	(+) Personal satisfaction in using renewable energy		(+) Creates jobs, stimulates local industry
Synthesis (the most telling facts)	(+) Satisfaction in using renewable, low-carbon power and in meeting carbon-reduction commitments (−) Dissatisfaction in increased energy bills to subsidize wind power.	(+) Generally positive impact on the environment. Off-shore wind farms avoid some of the intrusion of those on-shore. Re-use or recycling of critical elements should be straightforward.	(−) Power from wind farms not, at present, economic

concerns – a more considered response and exploration of mitigating measures (ultrasound, perhaps) is a better way forward.

The beauty of the countryside is a component of natural capital. All power-generating plant occupies space and is visually intrusive. The problem with wind farms is the scale of this intrusion if they are to contribute significantly to national needs for power. The long-term impact of acoustic intrusion is not known.

Manufactured and Financial Capital. The typical design-life of a wind turbine is 25 years. Building 50,000 turbines per year is a significant investment in energy infrastructure. Is it a good investment? Some argue that it is not because, without a subsidy, the electricity they produce is more expensive than that from gas-fired power stations. Governments have been inconsistent in dealing with subsidies, encouraging investment at one moment and then cutting the subsidy with little warning the next. Much will depend on the price and predictability of hydrocarbon fuels over the next 25 years and the cost of carbon-induced climate change.

Human and Social Capital. On the positive side, large-scale deployment of wind farms creates employment. If these jobs and the wealth they generate are distributed in a fair and equitable way, a contribution is made to Human Capital. The reduction in emissions is a contribution to a healthier population. The mix of energy sources increases independence and a distributed rather than a centralized power system is more robust, harder to disrupt and less vulnerable to a single catastrophic event.

On the negative side, the visual and acoustic intrusion, already mentioned, represents to many people a significant loss of quality of life. Schemes to re-invest a proportion of the revenues generated by the wind farm in the local community in ways that help everyone, coupled with research to reduce the acoustic problem, offer a way forward.

9.6 REFLECTION ON ALTERNATIVES

Short term The Prime Objective of wind farms – to generate electrical power with a low carbon footprint – appears to be met, making a contribution to Natural Capital. The relatively small scale of wind farms at present means that sites can be found for them without major disruption, and the reduction in emissions is a positive contribution to Human Capital. It is less clear that wind farms are economic, leaving a question mark over impact on Manufactured and Financial Capital.

Long term. Energy is one of mankind's most basic needs and electrical energy is the most versatile and valuable form it takes. We are in transition from a carbon-powered economy to one powered in other ways but the detailed shape of the future is not yet clear. A distributed energy-mix in the economy is desirable. If the cost of fossil fuels continues to rise in the future as it has in the past the economic case for the farms becomes stronger, but if grid-scale energy storage becomes necessary to smooth intermittent power from wind the cost again rises. Interestingly, electric vehicles, if they become the norm, might partly solve this problem. On average a private car is used for less than 4% of the day; the rest of the day is available for charging. Introducing intelligent battery charging that draws on power when there is surplus generating capacity turns the grid itself into a virtual storage device.

The evidence suggests that wind farms can make a contribution to national power needs but that it is likely to remain small. The intrusion caused by farms on a scale that could provide, say, half the nation's power appears to present very great problems. For now the dominant source power continues to be fossil fuels. Windfarms can offer one, perhaps transient, contribution while striving for other ways to establish a supply of clean energy and manage demand more effectively.

9.7 RELATED PROJECTS

These projects draw on some of the information assembled in the Wind farm analysis, but the prime objective, the stakeholders, the operating requirements and expectations of users all differ.

P9.1 **Wind power in Germany**. Germany has phased out nuclear power. The scope for expanding hydro-electric generation in the country is very limited. This places wind and solar photovoltaics as the two dominant options for generating low-carbon power. Explore the sustainability aspects of providing

50% of Germany's electrical power needs by wind turbines, bearing in mind that such a high dependence on an intermittent source will require grid-scale energy storage facilities.

P9.2 **Wind power obligations.** Research the Government obligations to expanding wind power in your country. How does it compare with that of the UK? Is it a sensible route for your country to take? Would solar PV be better? Use the information in this and the Solar Power case study as a short cut to some of the facts. But use all 5 steps of the method to develop your case.

Further Reading

Ardente F, Beccali M, Cellura M, Lo Brano V: Energy performances and life cycle assessment of an Italian wind farm, *Renewable Sustainable Energy Rev.* 12(1):200–217, 2008. http://dx.doi.org/10.1016/j.rser.2006.05.013.

Hood CF: *The History of Carbon Fibre*, www.carbon-fiber-hood.net/cf-history, 2009 (accessed 08/10).

Crawford RH: Life cycle energy and greenhouse emissions analysis of wind turbines and the effect of size on energy yield, University of Melbourne, Australia, *Renewable Sustainable Energy Rev.* 13(9):2653–2660, 2009.

Danish wind industry association: *What does a Wind Turbine Cost?* http://guidedtour.windpower.org, 2003 (accessed 08/10).

EWEA: *Costs and Prices, Wind Energy – The Facts*, vol. 2. www.ewea.org, 2003 (accessed 08/10).

Hau E: *Wind Turbines: Fundamentals, Technologies, Application, Economics*, ed 2, Springer, 2006, ISBN: 3-540-24240-6.

Hayman B, Wedel-Heinen J, Brondsted P: Materials challenges in present and future wind energy, *MRS Bull.* 33(4), 2008.

Martinez E, Sanz F, Pellegrini S, Jimenez E, Blanco J: *Life Cycle Assessment of a Multi-Megawatt Wind Turbine*, Spain, 2007, University of La Rioja. www.assemblywales.org (accessed 08/10).

McCulloch M, Raynolds M, Laurie M: *Life Cycle Value Assessment of a Wind Turbine*, Pembina Institute, 2000. http://pubs.pembina.org/reports/windlcva.pdf (accessed 08/10).

Musgrove P: *Wind Power*, Cambridge University Press, 2010, ISBN: 978-0-521-74763-9.

Nalukowe BB, Liu J, Damien W, Lukawski T: *Life Cycle Assessment of a Wind Turbine.* http://www.infra.kth.se/fms/utbildning/lca/projects%202006/Group%2007%20(Wind%20turbine).pdf, 2006 (accessed 08/10).

Vestas: *V82-1,65 MW, Vestas Wind Turbine Brochure*, 2008.

Vindmølleindustrien: *The Energy Balance of Modern Wind Turbines*, Wind Power Note, 16. http://www.apere.org/manager/docnum/doc/doc1249_971216_wind.fiche37.pdf, 1997, July (accessed 08/11).

Weinzettel J, Reenaas M, Solli C, Hertwich EG: Life cycle assessment of a floating offshore wind turbine, *Renewable Energy* 34:742–747, 2009. Elsevier Renewable Energy.

Windustry: *How much do wind turbines cost?* Windustry and Great Plains Windustry Project, Minneapolis. www.windustry.org/how-much-do-wind-turbines-cost, 2010 (accessed 08/10).

CHAPTER 10

Case Study: Electric Cars

Chapter Outline

10.1 Introduction and
    Background 151
10.2 Prime Objective and Scale 153
10.3 Stakeholders and Their
    Concerns 153
10.4 Fact-Finding 155

10.5 Synthesis with the Three
    Capitals 161
10.6 Reflection on Alternatives 163
10.7 Related Projects 165
Further Reading 166

10.1 INTRODUCTION AND BACKGROUND

The global production of cars in 2011 was 60 million unitsperyear, growing at 3.3% peryear. Cars account for 74% of production of motor vehicles and at present are responsible for about 20% of all the carbon released into the atmosphere[1]. National governments implement policies to reduce this source of emissions through taxation and incentives. One of the incentives is to subsidise electric vehicles (EVs) (Figure 10.1).

From a materials point of view, the major differences between electric and internal combustion (IC) cars are the replacement of the IC engine with electric motors that, at present, use neodymium–boron permanent magnets and the replacement of gasoline or diesel fuel by batteries (Figure 10.1). It is estimated that the global production of electric cars – either hybrids, plug-in hybrids, or fully electric – will exceed 16 million peryear in 2021 and will account

---

[1]www.epa.gov/climatechange/ghgemissions/sources.html.

Materials and Sustainable Development. http://dx.doi.org/10.1016/B978-0-08-100176-9.00010-4
Copyright © 2016 Elsevier Ltd. All rights reserved.

FIGURE 10.1
The Nissan Leaf, an electric vehicle (EV). Makers claim 0 grams CO_2/km.

for 20% of all vehicles manufactured[2]. EVs, particularly, are seen as the way to decarbonise road transport. France, Germany and the UK all have target EV sales of around 10% of all car sales by 2020 with the aim of reducing carbon emissions. Is this a realistically achievable sustainable development on a global scale?

10.1.1   Background Information

- Today's electric cars have 16 kWh batteries and a claimed range of up to 100 km between charges.
- An EV with this range requires about 1.5 kg of neodymium for the motors[3] and 7.3 kg of lithium, (equating to 0.46 kg lithium per nominal kWh) for the rechargeable batteries[4].
- The at-wheel energy required to propel a small car is between 0.6 and 1.0 MJ/km (0.17 and 0.3 kWhr/km)[5].
- Delivered electric power from a gas-fired power station has a carbon footprint of 500 g/kWh, or 140 g/MJ[6]; while from a coal-fired power station has larger carbon footprint.

Resources in the Appendix. Tables A.6 and A.17 contain data for neodymium and lithium. Tables A.9 and A.10 give the energy intensity and carbon footprint of fuels and electrical energy. Table A.13 lists the characteristics of batteries. Tables A.14 and A.16 give the energy demands and carbon emissions of transport systems.

---

[2]http://imsresearch.com/news-events/press-template.php?pr_id=2135.
[3]www.reuters.com/article/2009/08/31/us-mining-toyota-idUSTRE57U02B20090831.
[4]Tahil W, (2010) "How Much Lithium does a Li-Ion EV battery really need?.www.meridian-int-res.com and http://www.google.co.uk/search?sourceid=navclient&ie=UTF-8&rlz=1T4ADBR_enGB321GB323&q=how+much+lithium+is+in+a+battery.
[5]Telens Peiro L., Villalba Mendez G., and Ayres R.U. (2013) "Lithium: sources, production, uses and recovery outlook," JOM, vol. 65, pp. 896–996.
[6]See, for example, www.defra.gov.uk/publications/files/pb13773-ghg-conversion-factors-2012.pdf Table 3(c).

10.2   PRIME OBJECTIVE AND SCALE

Prime Objective and Scale are defined in the project statement. The prime objective is the de-carbonization of road transport. The scale is large (it has to be to make any significant difference to carbon emissions) – 10% of existing car production globally, equating to 8 million cars per year in 2020.

10.3   STAKEHOLDERS AND THEIR CONCERNS

As with wind farms, the national press reports the views of government, industry and the public about electric cars. Here are seven examples

- In his 2011 State of the Union address, widely reported, President Obama called for putting 1.2 million EVs on the road by 2015. This equates to 10% of the annual car sales in the United States.
- *Bloomberg Endorses Preparing Parking Spaces for E.V. Charging* (The New York Times, 14 February 2013). The mayor says he wants New York City to be a "national leader" in EVs.
- *Benefits of owning an EV. The UK Government offers a Plug-in EV Grant of 25% of the vehicle cost. EV's incur no road tax and are eligible for a 100% discount on the London Congestion Charge. Some London boroughs offer free parking for EVs.* (Transport for London web site, 2014)[7]
- *That Tesla Data: What It Says and What It Doesn't.* (The New York Times, 14 February 2013). The New York Times reporter responsible for covering energy, environment and climate change discovers the hard way that the claimed range of electric cars is sometimes a little overstated.
- CO_2 *emissions 0 g/km.* (The London Times, 24 February 2013). An advertisement for Nissan Leaf.
- *Are electric cars bad for the environment?* (The Guardian 4 February 2013). Norwegian academics argue that electric cars can be more polluting than claimed[8].
- *Leaf stalls* (The London Times 5 March 2013). Nissan admits that customers hesitate to buy its Leaf EV because of price and range anxiety.

---

[7]https://www.tfl.gov.uk/modes/driving/electric-vehicles.
[8]http://www.bbc.co.uk/news/magazine-22001356.

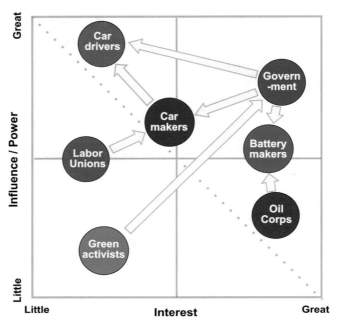

FIGURE 10.2
Stakeholder interest and influence.

- *Biofuels could cut CO$_2$ 'cheaper than electric cars'* – Businessgreen. com report the conclusion of a new (2013) report commissioned by oil giant BP, which part-owns the Vivergo ethanol plant[9].

These reports together with those quoted in Chapter 5 give an idea of the controversy surrounding EVs. They also give an insight into the stakeholders and their concerns (Figure 10.2). Among them are few listed below:

- *National Governments* encourage the take-up of electric cars in order to meet carbon-reduction targets and to reduce dependence (where it exists) on imported hydrocarbons.
- *Local city or state government* foresees pressure to provide charging points and specialized recycling facilities, particularly for battery materials.
- *Car makers and their suppliers* seek consistency of government policy to support a market for electric cars and a secure source for essential materials. They are uncertain of public acceptance of electric cars, making investment decisions difficult.

[9]http://www.businessgreen.com/bg/news/2295231/report-biofuels-could-cut-co2-cheaper-than-electric-cars.

- *Battery makers* seek to establish secure supply chains for the raw materials of the batteries, which include lithium and rare earths elements.
- *Mineral resource producing nations* are in a position to control materials supply chain and may wish to protect domestic car and battery makers rather than supply competitors with raw materials.
- *Oil companies* wish to retain share in the provision of fuels for future transport systems.
- *Labor Unions* are concerned about job creation, stable employment and improved pay and working conditions in the automobile sector.
- *Automobile associations and the driving public* share concerns about the range, battery life and replacement cost, and depreciation of electric cars.
- *Green Campaigners* lobby in favor of electric cars because of their concerns about the impact of gasoline and diesel-powered cars on the environment.

SUMMARY OF THE SIGNIFICANT CONCERNS

- That electric cars do not reduce the carbon footprint of transport.
- That the electric car market is not viable without government subsidies.
- That the supply chain for critical materials required for electric cars may be disrupted.
- That the range of electric cars is inadequate.
- That there are not enough charging points.
- That electric cars may impact the job market and demand for skills.

10.4   FACT-FINDING

Step 3

What information is needed to support or refute the claims made for EVs and the concerns expressed about them? What additional facts do we need for a rational discussion of the prime objective – 10% of cars fully electric by 2020? These questions are explored in the sections below. Figure 10.3 gives a overview.

Materials. The supply chain and availability of neodymium (Nd) was examined in the wind farm case study, Chapter 9. As noted there, the present annual global production of rare earths metals

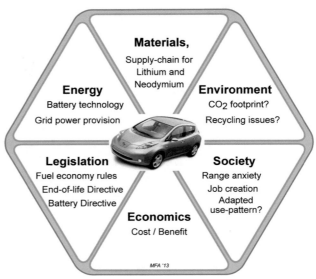

FIGURE 10.3
A fact-finding summary.

is about 134,000 tonnes per year, of which 15% (20,000 tonnes), on average, is neodymium. Over 95% of supply is from a single nation. The envisaged production of 16 million electric cars per year, each containing 1.5 kg of Nd would require, using today's technology

$$1.5 \times 8,000,000 \approx 12 \times 10^6 \text{kg} \approx 12,000 \text{ tonnes Nd/year}.$$

This is 60% of current global production, much of it already committed to other sectors. We already saw that there are no good substitutes for Nd-based magnets, so the constrained supply chain is a concern.

News-clip: material efficiency. *Global lithium deposits enough to meet electric car demands.* The world has enough lithium resources to power EVs for the rest of the century, according to Professor Greg Keoleian of the University of Michigan. "Responsible use of the resource is essential, even with abundant supplies. The key is to use lithium efficiently and not let it leak out of the economy after use. Expected demand for lithium has turned the element into a so-called critical material, with lawmakers on Capitol Hill working to secure US supply." *The New York Times, 28 July 2011.*

The other element of interest here is lithium (Li). Table 10.1 lists the main producers. The annual world production of lithium at present stands at 34,000 tonnes per year. The supply chain of Li is more diverse than that of Nd – 67% comes from Chile and Australia, the rest from a range of other nations.

The envisaged production of 16 million EVs per year, each with 16 kWh battery pack requiring 7.3 kg of Li, would, if battery design is unchanged, require

$$7.3 \times 8,000,000 \approx 58 \times 10^6 \text{kg} \approx 58,000 \text{ tonnes of Li/year}$$

or 170% of current world production. If car range is extended to meet consumer concerns the demand would be higher.

Are other battery technologies possible? To answer this we need information about energy-storage systems. That comes next.

Energy. Batteries are heavy. Weight is minimized by selecting the battery with the highest energy density. Figure 10.4 plots the energy density for energy-storage systems[10]. Lithium-ion batteries out perform all other battery types, although their energy

Table 10.1	Lithium Producing Nations (2011)
Nation	**Tonnes/year**
Chile	12,600
Australia	11,300
China	5,200
Bolivia	5,000
Argentina	3,200
Portugal	820
Zimbabwe	470
Brazil	160
World	**34,000**

Minerals.usgs.gov/minerals/pubs/commodity

[10]MacKay, D.J.C. (2009) "Sustainable energy – without the hot air" UIT Press, Cambridge, UK and Ashby, M.F. (2013) "Materials and the Environment" Butterworth Heinemann, Oxford.

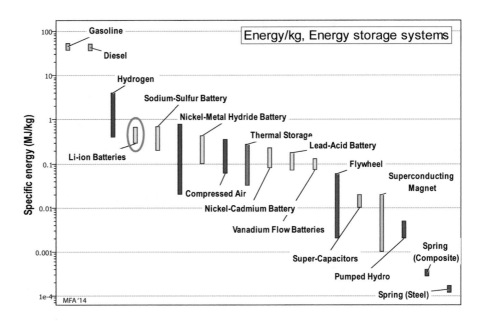

FIGURE 10.4
The specific energies of alternative energy storage systems using the energy storage data-table.

density, 0.6 MJ/kg, is still a factor 75 less than that of gasoline or diesel fuel (45 MJ/kg).

The at-wheel energy required to propel a small car is about 0.6 MJ/km. Thus the battery weight per unit range is roughly 1 kg/km. An acceptable range of 500 km (300 miles) would need a battery weighing half a tonne and costing, at today's prices, about $50,000.

There are about 1 billion cars on the world's roads. If 10% of these were EVs, driven 17,000 km (10,000 miles) per year, each consuming 0.6 MJ/km, they would draw

$$10^8 \times 0.6 \times 17,000 = 10^{12} \text{MJ/year}$$

from the national grid. An average power station produces 4×10^{10} MJ/year, so 23 additional power stations would be required to charge the cars. A country the size of the UK, France or Germany would require at least one additional power station to cope.

The Environment – can the Prime Objective be met? Electric cars will be charged from the national grid. Consider the carbon footprint of the car if the grid is largely fed (as in many nations it is) by gas-fired power stations. Delivered electric power from such stations has a carbon footprint of 500 g/kWh, or 140 g/MJ[11]. The energy in the form of gasoline or oil required to propel an efficient small car is about 2 MJ/km[12]. The conversion efficiency from gasoline to crankshaft power is at best 1/3, so for equivalent performance the electric motor replacing the IC engine must deliver about 0.6 MJ/km. The combined efficiency of a lithium-ion battery/electric motor set is at best 85% when the recharge cycle is included, so electrical energy of $0.6/0.85 \approx 0.7$ MJ/km must be provided from the grid. This carries a carbon penalty of

$$140 \times 0.7 \approx 100 \text{ grams per km.}$$

The median carbon emission of today's cars is about 160 g/km, but a number of contemporary models already emit less than 100 g/km. Thus until the grid is de-carbonized, carbon emissions from electric cars are no lower than those from an efficient gasoline or diesel powered vehicle. Power predominantly from nuclear sources (as in France) or from renewable sources (Norway, Iceland) changes the equation.

Legislation and Regulation. A search for legislation relating to private vehicles retrieves a number of European Directives and US Department of the Environment Acts:

European legislation

- EU Automotive Fuel Economy Policy on carbon emissions
- Fuel taxes
- EU Battery Directive
- End-of-Life Vehicles Directive (ELV)

US legislation

- CAFE rules
- Fuel taxes

All have a bearing on the viability of electric cars. We highlight one: the EU Battery Directive forbids the dumping of batteries in

---

[11]See the note 6 above.
[12]An efficient small car does about 16 km/litre of gasoline. One liter of gasoline has an energy content of 35 MJ/liter.

landfill; all must be recycled. Infrastructure for recycling Li-ion batteries on a large scale does not yet exist (3% of lithium-ion batteries are at present recycled[13]).

Economics[14] Batteries for electric cars are still very expensive – as much as $10,000 to $15,000[15], or one-third of the price of the vehicle – and can provide only limited range. The price of lithium-ion batteries fell during the 1990s but flattened out at about $600 per kWh. With fuel at $4/gallon (~$1/liter) in the United States and about $1.8/liter in Europe, the economics of electric cars looks unattractive. However a 2012 analysis carried out by McKinsey & Co[16] predicts that the price for lithium-ion batteries could fall by as much as two-thirds by 2020, down to around $200 per kWh. This, coupled with rising fuel price, might tip the balance.

A current economic concern is the investment in recharging points: providers are waiting for the number of electric car drivers to rise but drivers are waiting the number of stations to rise. Some governments are willing to subsidize charging points, but mainly in cities.

Society. Automobiles give independence. Their manufacture creates employment. They also occupy space and, in conventional form, are responsible for noise and emissions. Secondary benefits of the electric car include reduction in noise and the ability to confine carbon release to power stations where it can be handled more effectively.

The cost, the limited range and absence of charging points for electric cars impedes their acceptance at present. Governments recognise these as problems and seek to reduce their impact by subsidies on EV purchase and installing and subsidising charging points.

---

[13]See the note 5 above.
[14]The Washington Post, http://www.washingtonpost.com (accessed 02.04.2013).
[15]The Wall Street Journal, http://online.wsj.com/article/SB10001424052702304432704577350052534072994.html (accessed 17.04.12).
[16]McKinsey, July 2012. http://www.mckinsey.com/insights/energy_resources_materials/battery_technology_charges_ahead.

SUMMARY OF SIGNIFICANT FACTS

- The supply-chain for neodymium and lithium is at present inadequate to support making 8million electric cars per year. Alternatives carry a performance penalty.
- If charged from a national grid fed by gas or coal-fired power stations the carbon footprint of the car is at least 100 grams CO_2/km.
- The weight and cost of batteries limits the range to less than about 160 km per charge.
- Sales of electric cars at present depend on government subsidies of up to 20% of the price of the car.
- Legislation requires that 85% of the car, including the batteries be recycled. Facilities for recycling lithium-ion batteries and neodymium magnets are not, at present, in place.

10.5   SYNTHESIS WITH THE THREE CAPITALS

What, then, is the likely impact of wide use of electric cars on the three capitals? These are questions for debate, informed by the data generated by the fact-finding step. Here is one view for discussion, summarized in Table 10.2.

Natural Capital. EVs that use today's technology rely on at least two "critical" elements: neodymium and lithium. The analysis of demand created by EVs and the distribution of source nations for these elements were not reassuring: The projected demand for neodymium for cars in 2020 is about half the current (2011) global production, most of it coming from a single nation. Some of the demand in 2021 could be filled by recycling, not at present practiced. The design life of an electric car is of order 12 years. If the vehicles are leased, so that large groups of them are managed by a single enterprise, the recovery, reconditioning or recycling at end of life is straightforward. If they are sold, as cars are now, to individual purchasers then collection for recycling becomes more difficult but still manageable. A similar exploration for lithium indicated a broader supply base but a demand in 2020 that exceeds current production capacity. These facts point to a technology that makes large demands on critical elements with inadequate current supply.

Table 10.2   Synthesis: Influence of the Facts on the Three Capitals for Electric Cars

	Human and social capital – People *Health? Wellbeing? Convenience? Culture? Tradition? Associations? Perceptions? Contributes to equality? Morality?*	Natural capital – Planet *Can prime objective be met? Are stakeholder concerns addressed? Are there unwanted consequences?*	Manufactured capital – Prosperity *Cost-Benefit? (Cost facts vs. Eco facts) Legitimacy? (Conformity with law)*
Materials	(-) Creates dependence on rare-earth and lithium-producing nations	(-) Creates demand for critical elements, notably lithium and neodymium, in kg-quantities per car (+) Use of Li and Nd in kilogram-scale components makes collection for recycling easier	(-) Requirement to create recycling infrastructure for lithium- and rare-earth elements
Energy	(+) Coufuels in oil-poor nations ld reduce dependence of imported fossil	(-) Very little contribution to carbon emissions unless national grid is decarbonized	(+) Creates employment in energy sector (-) Need for additional power stations (-) Need for investment in recharging points
Environment	(+) Reduces emission levels in large cities	(+) Offers potential for clean energy for transport	–
Legislation	(+) Helps meet the nation's commitments to reduce emissions	(+) Take-back and recycling legislation reduces waste stream, contribute to a circular economy	(-) Meeting end-of-life regulations creates additional costs
Economics	(-) Need to subsidize sales of electric cars becomes a "green" tax.		(-) Profitability uncertain without government subsidies (-) Large capital investment in new electricity generating plant to provide for charging
Society	(+) Satisfaction in using an "emission-free" transport (-) Range anxiety (-) Paucity of charging points		(+) Creates employment in high-tech industry (+) Creates jobs, stimulates local industry
Synthesis (the most telling facts)	(+) Satisfaction reducing environmental impact (-) Dissatisfaction with green taxes used to subsidize green transport	(+) Ultimate success dependent on new battery technology, decarbonized grid and adaptation to range limitations – impossible in short term (+) But potentially possible in the long term.	(-) Electric cars not, at present, economic. Many issues to be resolved to make it so.

Does the all-electric car achieve its prime objective, that of helping to de-carbonize road transport? The carbon footprint of the electric car, if charged from the national grid of a typical Western nation, is roughly 100 grams per km. An increasing number of small IC driven cars already do better than this. We conclude that the prime objective is not achieved until the national grid is itself de-carbonized or an independent low-carbon source of electrical power is available. Neither appears achievable in the short (6 year) term.

Manufactured Capital. The aim of 8 million EVs per year by 2020, using today's technology, is achievable only if three conditions are met: the supply chain for the critical elements on which they depend is expanded and given a broader base; provision for recycling these elements is established; and grid-electricity generation capacity is increased.

Creating plant to build more than a million electric cars per year is a large investment in manufacturing technology. Is it a good investment? Some argue that it is not because, like wind-turbines, EVs are not competitive in cost without a government subsidy. As with all energy-using products the unknown is the price of hydrocarbon fuels over the next 20 years and the currently externalised cost of carbon-induced climate change.

Human Capital. A healthy manufacturing base makes a positive contribution to human capital: the jobs created by the automobile industry contribute to wealth and potentially to the well-being of the population of the nation in which they are built. But EVs can contribute to human capital in this way only if they are widely accepted by the driving public. The limited range, at present, is an obstacle to acceptance.

10.6   REFLECTION ON ALTERNATIVES
10.6.1   Short Term

This is the moment to consider alternatives. Can the prime objective be achieved in the way assumed in the remit – by replacing petrol-driven cars by EVs that are used in the same way? It does not seem so. EVs cannot provide the range, convenience of refuelling or (at present) the economy that consumers expect. Even more telling: charging EVs from the national grid of most nations carries a carbon footprint larger than that of many small IC and hybrid-powered cars today.

10.6.2   Long Term

To expose innovation potential we must return to the "society" and "economics" dimensions and abandon the idea of an electric car as a straight substitute for one with an IC engine.

Electric cars are good for short journeys. Could the public be re-educated to think of electric cars in a new way, not as a simple replacement for an IC engine car but as a vehicle well-adapted for urban use, when range is less important? Could it be made attractive to own a small electric car for daily commuting and rent a larger IC car for longer journeys, vacations or employment that required one? Or could large companies provide electric cars and on-site charge-points for staff, subsidising their commuting in a way that best used EVs? A shift from private ownership to fleet ownership by municipalities, service providers and employers with provision of recharging points at supermarkets, car-parks and place of work could make better use of the strengths of electric transport.

A central issue for electric transport is that of energy density. Suppose we accept that transport is best powered by high energy-density fuels with which batteries cannot compete. Technology exists for synthesizing hydrocarbons from CO_2. Rather than using electrical power to charge batteries, could it be used to synthesize methanol or ethanol to drive efficient IC-powered cars? The infra-structure for fuel distribution and maintenance already exits, and by drawing the CO_2 from carbon-intensive industries such as power-stations, or cement works or simply from the atmosphere, true carbon-neutrality might be possible.

EVs could perform another, quite different function, that of making intermittent renewable energy from wind and solar sources more practical. Most cars are in use for less than 4% of the average day. EVs can then be charged during off-peak hours at cheaper rates while helping to absorb excess night time generation. The excess rechargeable battery capacity can then provide power to the electric grid in response to peak load demands. The vehicles serve as a distributed battery storage system to buffer intermittent power generation.

10.7   RELATED PROJECTS

These projects draw on some of the information assembled in the electric car analysis, but the prime objective, the stakeholders, the operating requirements and expectations of users are all significantly different from those for electric cars.

P10.1 **Electric buses.** Repeat the analysis of this chapter, applying it instead to electric buses for inner city transport. The use-pattern of buses has the following characteristics. They follow regular routes. The operating company knows the length of each route, how frequently they are traversed and where each bus will be at each moment in the day. It has central service facilities and dedicated charging facilities. A privately owned car has none of these characteristics. How much difference does this make?

P10.2 **Electric bicycles.** An electric bicycle ("e-bike") is a bicycle with an integrated electric motor that assists or replaces pedaling. In most countries they are classified as bicycles and require no license or registration. They cost between €425 and €1800 ($500–$2100), are limited to about 30 km/hr and have a range of about 25 km. E-bikes are aimed at commuters – you still get some exercise but you also do not arrive at your destination so hot and sweaty. The take-up of electric cars was hindered by cost and range. Are electric bicycles a more sustainable option?

P10.3 **Cobalt in electric cars.** The lithium-ion battery of an electric car with a range of 160 km contains about 9.4 kg of cobalt. Making the same assumptions about scale and timing as those of the case study of this chapter, would you anticipate that cobalt supply might be a constraint?

Further Reading

Electric Vehicles: Charged with Potential, Royal Academy of Engineering, May 2010.

Cebon D, Collings N: Letter to the Editor, *Ingenia Magazine*, 5 September, 2010.

The Race to Build Really Cheap Cars, Bloomberg Business Week, April 23, 2007.

The Hydrogen Illusion, Cogeneration and on-site power production, March–April 2004. http://www.efcf.com/reports/E11.pdf

Mintz M, Folga S, Molburg J, Gillette J: Cost of Some Hydrogen Fuel Infrastructure Options, *Transportation Research Board*, 2002.

Assessment of technologies for improving light duty vehicle fuel economy, Committee on the Assessment of Technologies for Improving Light-Duty Vehicle Fuel Economy; National Research Council, ISBN: 978-0-309-15607-3, 260 pages, 2010.

CHAPTER 11

Lighting

Chapter Outline

11.1 Introduction and Background Information 167

11.2 Prime Objective and Scale 169

11.3 Stakeholders and Their Concerns 170

11.4 Fact-Finding 172

11.5 Synthesis with the Three Capitals 177

11.6 Reflection on Alternatives 179

11.7 Suggested Projects 179

Further Reading 180

11.1 INTRODUCTION AND BACKGROUND INFORMATION

Electric lighting is a recent invention. For thousands of years our ancestors lit their caves and, later, their buildings with oil lamps – a wick in a bowl of animal fat. The first solid-state lighting – the candle – was developed around CE 400, using beeswax instead of oil. Gas lighting appeared about 1810, enabling street and theatre as well as domestic lighting on a large scale (and causing many fires). We have Thomas Edison to thank for the electric light; he invented the incandescent filament lamp and founded

Materials and Sustainable Development. http://dx.doi.org/10.1016/B978-0-08-100176-9.00011-6
Copyright © 2016 Elsevier Ltd. All rights reserved.

the wonderfully named Edison Electric Illuminating Company (1874) to illuminate the citizens of New York, from where it spread to the rest of the world. Other innovators followed, but none with the impact on everyday life that Edison made. Fluorescent lamps became commercially available in 1938; light emitting diodes (LEDs) in 1965. Both are now available as alternatives to incandescent bulbs. And suddenly we need them. We have lived happily with the incandescent lamp for 140 years, but now low-key legislation in the US, the EU and elsewhere has banned them. Here are examples.

Directive 2009/125/EC of the European Parliament sets a program for the phasing out of incandescent lamps, requiring their total replacement by more energy-efficient alternatives by March 2014, with a minimum bulb life of 6000 h. The US (EISI 2007), Canada and many other nations have implemented similar legislation with a similar time scale.

Does this qualify as a sustainable development? In this case study we compare incandescent lamps (ILs) with their most-used replacement, compact fluorescent lamps (CFLs).

11.1.1   Background Information

The SI unit of luminous intensity is the candela, symbol *cd*. It is based on light of frequency of 540×10^{12} hertz, which is yellowish-green. Candela is a measurement of light at source but does not tell us how powerful the light is some distance away from the source. The light illuminating a unit surface area is called the luminous flux. The SI unit of luminous flux is the lumen, symbol *lm*. One *lm* is the amount of light emitted in one second in a unit solid angle of one steradian by a source of one candela. (A steradian is the SI unit of solid angle, equal to the angle at the center of a sphere subtended by a part of the surface equal in area to the square of the radius.) The SI unit of illuminance (luminous flux per unit area) is the lux, symbol *lx*, so $1 lx = 1 lm/m^2$. The difference between the lumen and the lux is that the lux takes into account the area over which the luminous flux is spread. A flux of 1000 lumens, concentrated into an area of $1 m^2$, lights up that square meter with an illuminance of 1000 lux. The same 1000 lumens, spread out over $10 m^2$, produce

Source of Light	Luminous Efficiency Lumens/Watt (%)	Claimed Lifetime (h)	Approximate Price (US$)
Candle	0.3 (0.04%)	1–2	0.2–2
Gas light	1–2 (0.15–0.3%)	500–1000	–
Incandescent (IL)	13–19 (2–3%)	1000–2500	0.5–1.0
Compact fluorescent (CFL)	45–75 (8–11%)	5000–10,000	3–5

Table 11.1 **Light Sources – Approximate Efficiencies, Lifetimes and Prices**

a lower illuminance of only 100 lux. The sun's illumination in summer can reach 130,000 lux; at night the light of the moon is less than 1 lux. A typical living room has an illuminance of about 50 lux.

If electricity is converted into the yellowish-green light with 100% efficiency it produces 683 lm/W. Even the most efficient light bulbs reach only 15% of this value; incandescents barely exceed 3% (Table 11.1).

Resources in the Appendix. Tables A.2 and A.17 contain data for world production and environmental characteristics of elements. Table A.10 gives the energy intensity and carbon footprint and electrical power generation. Table A.15 lists various conversion efficiencies, including that of electrical power to illumination.

11.2 PRIME OBJECTIVE AND SCALE

Step 1

Prime objective, size and time scale. The prime objective in banning incandescent light bulbs is to save energy by forcing, globally, a move to alternative low energy bulbs. About 2 billion light bulbs are sold in the US each year[1]. Scaling this to global sales, allowing for lower consumption in less developed nations, suggests global sales of around 20 billion per year. The time scale is set by the remaining life of the existing incandescents, which we take to be approximately 2 years.

[1]US Department of Energy Statistics, May 2012.

11.3   STAKEHOLDERS AND THEIR CONCERNS

Step 2

Here are contemporary news quotes about lighting.

- *Lights out for the incandescent light bulb as of Jan. 1, 2014. Melissa Andresko, communications director for lighting company Lutron, reminds us that incandescent bulbs are now banned*[2] (Fox News, 31 December, 2013).
- *Fluorescent light bulbs emit high levels of UV radiation. Researchers at Stony Brook University found energy-efficient bulbs emit harmful ultraviolet radiation*[3] (CBS Los Angeles, 18 October, 2012).
- *Understanding the Rare Earth Crisis. Rare earth oxides of lanthanum, cerium, terbium, yttrium and europium are used to make white light from the UV radiation in fluorescent lamps. Today, no viable substitutes exist for rare earths in lamps*[4] (Report from GE Lighting, 2014).
- *Unrecycled CF light bulbs release mercury into the environment. Only about 2% of residential consumers and one-third of businesses recycle CFLs, according to the Assn. of Lighting and Mercury Recyclers. As a result, U.S. landfills are releasing more than 4 tons of mercury annually into the atmosphere and storm water runoff, a study says* (Los Angeles Times, 7 April, 2011).
- *Basking in a New Glow. As incandescent bulbs disappear and compact fluorescents grow dim, the stage is set for better, cheaper LEDs that last much longer* (New York Times, 13 February, 2014).
- *The great LED light bulb rip-off: One in four expensive 'long-life' bulbs doesn't last anything like as long as the makers claim. Some did not even reach the EU's new minimum of 6000 h which comes into force in March*[5] (The Daily Mail, 26 January 2014).

These quotes suggest the key stakeholders (Figure 11.1).

- Government, who see banning incandescents as a way of saving energy and stimulating development of alternatives.
- Bulb makers, who perceive CFLs as a lucrative market.

---

[2]http://www.foxnews.com/tech/2013/12/31/end-road-for-incandescent-light-bulb/.
[3]http://losangeles.cbslocal.com/.
[4]http://www.gelighting.com/LightingWeb/na/images/GE_Rare_Earth_WhitePaper.pdf.
[5]http://www.dailymail.co.uk/news/article-2546363/.

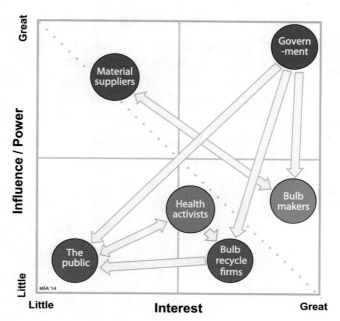

FIGURE 11.1
Stakeholders and influence.

- The public, who are concerned about the quality of light and health risks associated with alternatives to incandescent lighting but are careless in disposing of CFLs at end of life.
- Bulb recyclers, who find CFLs difficult to recycle.
- Environmentalists, concerned about toxic legacy of CFLs.
- Material suppliers, concerned with volatile pricing of rare earths oxides.

SUMMARY OF THE SIGNIFICANT CONCERNS

- Will a switch from incandescent to alternative light bulbs save energy?
- Do CFLs or LEDs pose health risk from radiation, hazardous chemicals, flicker or poor light quality?
- Is the supply of critical elements essential for the manufacture of CFLs at risk?
- Can disposal be managed in a way that avoids the build-up of toxic waste?

These concerns flag the targets for research in the fact-finding stage that follows.

11.4   FACT-FINDING

Step 3

What information is needed to establish whether CFLs meet both the prime objective and the stakeholder concerns? We will calculate the material consumption, energy, global warming potential and cost of providing a fixed illumination, 800 lumens, for 1000 h, comparing compact fluorescent with incandescent lamps.

Figure 11.2 summarizes the issues needing research. No judgements at this point, only facts.

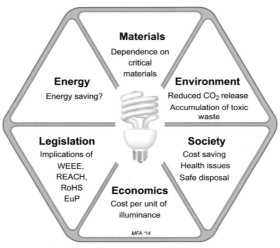

FIGURE 11.2
A fact-finding summary.

1. **Material attributes**[6]. Incandescent bulbs contain few materials. CFLs contain far more, many of them classified as critical[7]. Table 11.2 compares ILs with CFLs with the same illuminating power.

---

[6]OSRAM Life Cycle Analysis of CFL. http://www.Osram.com/Osram_com/sustainability/sustainable-products/life-cycle-analysis/lca-of-a-compact-fluorescent-lamp/index.jsp (accessed 25.04.2014.).
OSRAM Life Cycle Analysis of Incandescent. http://www.Osram.com/Osram_com/sustainability/sustainable-products/life-cycle-analysis/lca-of-an-incandescent-lamp/index.jsp (accessed 25.04.2014.).
OSRAM Life Cycle Analysis of LED. http://www.Osram.com/Osram_com/sustainability/sustainable-products/life-cycle-analysis/lca-of-an-led-lamp/index.jsp (accessed 25.04.2014.).
[7]US Department of Energy Critical Materials Strategy. http://energy.gov/sites/prod/files/DOE_CMS2011_FINAL_Full.pdf (accessed 02.05.2014.).

		Incandescent 60W	Compact Fluorescent, 13W
Bulb characteristics	Bulb weight (g)	33	50
Non-critical materials	Glass (g)	29	27
	Ferrous metals (g)	0.1	0.75
	Aluminum (g)	1.2	–
	Plastic and resin (g)	–	–
	Cement (g)	2.0	3.3
	Unspecified electronics (g)	–	–
Critical materials	Antimony (g)	–	0.025
	Arsenic (g)	–	–
	Gallium (g)	–	–
	Germanium (g)	–	–
	Lead (g)	–	0.25
	Mercury (g)	–	0.004
	Tungsten (g)	0.05	0.0025
	Molybdenum (g)	0.06	–
	Phosphors (g)	–	0.23
	Barium (g)	–	0.005
	Cerium (g)	–	0.01
	Europium (g)	–	0.005
	Lanthanum (g)	–	0.005
	Strontium (g)	–	0.005
	Yttrium (g)	–	0.1

Table 11.2   Bills of Materials for Lamp Bulbs

Critical material supply chain. The weight of materials in a bulb is small but their numbers are large. Both bulb types rely on materials classed as critical. The small quantity per bulb and the great dispersion because of the large numbers mean that recovery is impractical. Does the demand represent a significant drain on scarce resources?

Making 10 billion incandescent bulbs per year consumes roughly 500 tonnes of tungsten per year, none of it recoverable. The world production of tungsten is 72,000 tonnes per year[8], so bulbs dissipate about 0.5% of global production. The drain on molybdenum is about the same.

CFLs last longer than incandescent bulbs, so, if CFLs replace them, fewer will be needed. Suppose, as an example, an annual production of 1 billion. The anticipated demand for yttrium for the phosphors is then of order 100 tonnes per year. The annual production is 8900 tonnes per year, so CFLs, if adopted as the universal domestic light source, would disperse about 1% of global production.

2. **Energy**. Energy consumption for the two bulb types are compared in Table 11.3, scaling the embodied energy of the bulb by its life, and converting electrical energy of primary (oil-equivalent) energy with a conversion efficiency of 0.33. The energy of use exceeds that to make each bulb by a factor of around 100. CFLs consume about one-fifth as much energy per lumen as incandescents. The energy pay-back time in hours, t^* for the replacement of incandescent by CFLs is

$$t^* \times (\text{power, incandescent} - \text{power, CFL}) = (\text{embodied energy}, \text{CFL} - \text{embodied energy}, \text{ incandescent}$$

from which t* = 77 h.

3. **The Environmental properties**. The lower part of Table 11.3 gives equivalent data for the global warming potential. The lower energy consumption translates into a lower GWP for CFLs.

CFLs contain mercury. If dumped, 1 billion CFLs per year each containing 4 mg of mercury, result in 4 tonnes per year to landfill. CFLs emit some UV radiation and can flicker making it dangerous to use them with power tools.

4. **Legislation**. The design, manufacture and disposal of light bulbs must comply with a catalog of restrictions and legislation aimed at energy efficiency, reporting of hazardous materials and safe disposal at end of life. Table 11.4 gives examples. Incandescent bulbs do not contain hazardous

---

[8]USGS Mineral Commodity Summaries. http://minerals.usgs.gov/minerals/pubs/mcs/2014/mcs2014.pdf.

Table 11.3	Energy and Global Warming Potential to Provide 800 lumens for 1000 h		
		Incandescent[9]	Compact fluorescent[10]
Power to provide 800 lumens (W)		60	13
Claimed bulb life (h)		1000	8000
Embodied energy, bulb (MJ per unit)		3.2	16.3
Embodied energy, bulb, pro-rated to 1000 h (MJ)		3.2	2.0
Energy consumption, 1000 h use (MJ_{elec})		216	47
Energy consumption, 1000 h use ($MJ_{primary}$)[a]		713	154
Total energy, 800 lumens, 1000 h ($MJ_{primary}$)		**716**	**156**
GWP^b, bulb (kg per unit)		0.2	1.0
GWP pro-rated to 1000 h (kg)		0.2	0.13
GWP of use, 1000 h use (kg)		33	6.1
Total GWP, 800 lumens, 1000 h (kg)		**33.2**	**6.2**

[a]Assuming 1 MJ_{elec} requires 3.3 $MJ_{primary}$.
[b]GWP = Global warming potential = CO2,eq.

Table 11.4	Legislation Bearing on Light Bulbs
Sector	**Legislation**
Registration and control of hazardous materials and chemicals	EU Registration Evaluation & Authorization of Chemicals Directive (REACH)
	EU Restriction of Hazardous Substances Directive (RoHS)
	China REACH
	China Restriction of Hazardous Substances (RoHS)
Energy and Product design	EU Energy-using Products Directive (EuP)
	US Energy Independence and Security Act, 2007 (EISA 2007)
	France Grenelle 2 Regulations
End of life and Control of Waste	EU Waste Electrical and Electronic Equipment Directive (WEEE)
	Japanese Household Appliance Recycling Law (HARL)

[9]OSRAM Life Cycle Analysis of Incandescent, Scaled Up from 40 W to 60 W. http://www.Osram.com/Osram_com/sustainability/sustainable-products/life-cycle-analysis/lca-of-an-incandescent-lamp/index.jsp (accessed 25.04.2014.).
[10]OSRAM Life Cycle Analysis of CFL, Scaled Up from 8 W to 10 W. http://www.Osram.com/Osram_com/sustainability/sustainable-products/life-cycle-analysis/lca-of-a-compact-fluorescent-lamp/index.jsp http://www.ecy.wa.gov/programs/swfa/mrw/pdf/Presentations/Will%20perry%20Chemistry%20of%20Fluorscent%20lamps.ppt.pdf (accessed 25.04.2014.).

Table 11.5 Cost of Providing 800 lumens for 1000 h		
	Incandescent	**Compact Fluorescent**
Cost/bulb ($)	0.5	5
Power to provide 800 lumens (W)	60	13
Life (h)	1000	8000
Bulb cost, pro-rated to 1000 h ($)	0.5	0.6
Energy consumption, 1000 h (kWh)	60	13
Cost of electricity, 1000 h at $0.2/kWh ($)[a]	12	2.7
Total cost for 1000 h, 800 lumens ($)	**12.5**	**3.1**

[a]Cost of domestic electricity in Europe.

chemicals, but have succumbed to legislation relating to energy efficiency. CFLs contain hazardous chemicals (mercury, heavy metals, phosphors). Disposal is a present problem: despite legislation requiring safe disposal, most bulbs still end up in landfill. CFLs can be recycled but there are few facilities for doing so.

5. **Economics**. Table 11.5 gives a cost comparison of the two types of bulb. The initial cost of the bulb is pro-rated to 1000 h using the claimed bulb life. The cost of electric power over 1000 h assumes domestic electricity at $0.2/kWh (in the US it is less). In all three cases the initial cost of an incandescent bulb is small compared to the cost of the energy it uses over its lifetime. CFLs provide light at about one-quarter of the cost of incandescent bulbs.

6. **Society**. Gaining acceptance for new technology can be difficult. CFLs have a long warm-up time[11], a different quality of light and a flicker that some find unattractive. Overall there is a clear social benefit in their lower cost and emissions.

SUMMARY OF SIGNIFICANT FACTS

- Domestic lighting with compact fluorescent bulbs is more energy efficient, less expensive and generates less emissions than lighting with incandescent bulbs.

[11]Commission of the European Communities. Full Impact Assessment. http://ec.europa.eu/governance/impact/ia_carried_out/docs/ia_2009/sec_2009_0327_en.pdf (accessed 13.11.2013.).

- Both bulb types use critical materials in ways that are not, at present, recoverable.
- The consumption of critical materials in bulbs is not at present a significant fraction of world production.
- CFLs contain mercury that accumulates in land-fill if the bulbs are dumped.
- WEEE legislation requires that bulbs be recycled but in practice domestic bulbs seldom are.
- Resistance to adopting CFLs derives from the perception that they contain toxic substances and emit radiation.

11.5   SYNTHESIS WITH THE THREE CAPITALS

What impact do the facts have on the three capitals? Some impact on more than one capital; for simplicity, we assign the influence to the one most affected. As explained in Chapter 3, there is no single answer to this question; the weight given to the facts depends on values, culture, beliefs and ethics. Here we present one view. The main points are summarized in Table 11.6, color-coded and with positive (+) and negative (–) positive influences indicated.

Natural capital. Is the prime objective – that of reducing the energy required for domestic lighting – achieved by the enforced replacement of incandescent lights? Yes, and quickly. The energy break-even time for CFLs is less than 100 h and they offer reduced emissions. The demand they create for critical materials is a small fraction of global supply and concern about toxicity can be managed by proper disposal.

Manufactured and Financial capital. Fluorescent lighting is a disruptive technology. In a forced change from one technology to another there are both winners and losers. The change itself has to be managed in ways that conform to laws and regulations, meaning proper labelling, disposal and recycling, all of which require investment. But new markets with potentially higher value-added are created by the new technology.

Human and Social capital. Replacement of incandescents offers a significant economic advantage to the consumer. Concerns about health hazards are allayed by provision of proper disposal. Reduced emissions ultimately contribute to quality of life.

Table 11.6   Synthesis: Influence Replacing Incandescent Bulbs by CFLs on the Three Capitals

	Human and social capital – People *Health? Wellbeing? Convenience? Culture?* *Tradition? Associations? Perceptions?* *Equality? Morality?*	Natural capital – Planet *Can prime objective be met?* *Are stakeholder concerns addressed?* *Unwanted consequences?*	Manufactured capital – Prosperity *Cost – Benefit? (Cost facts vs. Eco facts)* *Legitimacy? Conformity with law?*
Materials	(–) CFLs contain toxic chemicals	(+) Long CFL life reduces need for replacement (–) Increased dependence on critical materials	(+/–) Fluorescent lighting is a disruptive technology – some winners, some losers
Energy		(+) Significant reduction in energy demand	
Environment		(+) Significant reduction in emissions	
Legislation	(+) Reduced emissions contributes to quality of life	(+) Complies with EuP legislation Recycling legislation reduces waste stream	(–) Complying with WEEE legislation requires investment in CFL disposal
Economics	(+) Significant reduction in cost of domestic lighting	–	(+) Advanced technology lighting represents higher value-added
Society	(–) Health concerns from UV radiation, flicker and changed quality of light (–) Apprehension at impact of new technology		(+) Energy labeling and government promotion stimulates acceptance
Synthesis (the most telling facts)	(+) Significant economic advantages to consumer (–) Misconceptions act as drag on acceptance	(+) Reduced energy demand (+) Reduced emissions (–) Dependence on critical materials	(+) Advancing technology creates new markets

11.6   REFLECTION ON ALTERNATIVES

Short term. A win–win proposition. The prime objective is achieved and the causes for concern are manageable.

Longer term. CFLs may be 5 times more efficient than incandescents, but they only raise the efficiency from 2% to 10%. So the new bulbs are a step in the right direction but there is still a long way to go. Further research and development of LED lighting will increase efficiency further. But if we wish to move towards a truly circular materials economy, safe disposal of bulbs at end of life is not enough; instead, ways of recovering the tiny quantities of critical materials in each bulb will be needed.

11.7   SUGGESTED PROJECTS

These projects draw on some of the information assembled in this analysis, but the stakeholders, the operating requirements and expectations of users may differ significantly.

P11.1 Make a comparison like the one in the case study for the introduction of LED lighting to replace CFLs. The table provides some background. A 10 W LED bulb has the same illuminating power as a 13 W CFL or a 60 W incandescent. Governments have not yet mandated the phase out of CFLs and there are no signs at present they will. Take up will therefore depend on consumer choice. What can LEDs offer that CFLs don't?

	Incandescent[12]	Compact fluorescent[13]	LED[14]
Power to provide 800 lumens (W)	60	13	10
Approx. price of bulb of this power ($)	0.5–1.0	3–5	20–40
Claimed bulb life (h)	1000	8000	25,000
Embodied energy, bulb (MJ per unit)	3.2	16.3	35
GWP, bulb (kg per unit)	0.2	1.0	2.4

P11.2 Conduct an analysis like that of this case study on a national scale for the country in which you live. How much energy is used for lighting? How much energy would be saved if all domestic lighting (or office lighting, or traffic signals) were converted to LED sources? Is there a will at the government level to make such a change? Who would be affected by it? What are their concerns – would they welcome or oppose the change? What legislation requires compliance? What provision exists for recycling bulbs?

P11.3 The long life of LED bulbs (25,000 h – perhaps 25 years of normal domestic use) is seen as an asset. But is it? The changes in lighting technology in the last 25 years have been dramatic – the incandescent bulbs of 25 years ago, standard then, are now banned. Will new technology in the next 25 years displace LEDs? Rework the economics table (Table 11.5) for LED bulbs assuming that the life before redundancy of an LED bulb is 5000 h, not 25,000 h. How much difference does it make?

Further Reading

MacKay DJC: *Sustainable energy – without the hot air*, Chapter 9. UIT Cambridge, 2008, ISBN: 978-0-9544529-3-3(Clear thinking and common sense applied to energy generation and use.).

[12]OSRAM Life Cycle Analysis of Incandescent, Scaled Up from 40 W to 60 W. http://www.Osram.com/Osram_com/sustainability/sustainable-products/life-cycle-analysis/lca-of-an-incandescent-lamp/index.jsp (accessed 25.04.2014.).

[13]OSRAM Life Cycle Analysis of CFL, Scaled Up from 8 W to 10 W. http://www.Osram.com/Osram_com/sustainability/sustainable-products/life-cycle-analysis/lca-of-a-compact-fluorescent-lamp/index.jsp http://www.ecy.wa.gov/programs/swfa/mrw/pdf/Presentations/Will%20perry%20Chemistry%20of%20Fluorscent%20lamps.ppt.pdf (accessed 25.04.2014.).

[14]OSRAM Life Cycle Analysis of LED. http://www.Osram.com/Osram_com/sustainability/sustainable-products/life-cycle-analysis/lca-of-an-led-lamp/index.jsp (accessed 25.04.2014.).

CHAPTER 12
Solar PV

Chapter Outline

12.1 Introduction and Background
      Information 182
12.2 Prime Objective and Scale 183
12.3 Stakeholders and
      Their Concerns 184
12.4 Fact-Finding 186

12.5 Synthesis with
      the Three Capitals 191
12.6 Reflection on Alternatives 192
12.7 Suggested Projects 194
Further Reading 195

12.1 INTRODUCTION AND BACKGROUND INFORMATION[1]

Present (2014) global electricity consumption is 21×10^{12} kWh (21,000 TWh) per year, equivalent to an average continuous power consumption of 2.4 TW. At present 80% of it is generated by burning fossil fuels, releasing about 10 billion tonnes of carbon emissions per year. Many nations have carbon-reduction targets; that in Europe, for example, is a 20% reduction by 2020 and an 80% reduction by 2050. That is more than a target; it is a commitment to a sustainable development.

12.1.1 Background Information

Governments seeking ways to meet carbon-reduction targets see electricity generation from renewable sources as part of the answer. Direct conversion of sunlight via photovoltaic (PV) panels has many attractions. Sunlight falls everywhere. Once installed, solar PV emits no carbon, it is silent and it reduces dependence on imported energy. Solar energy, however, is not shared equally between nations (Figure 12.1). Despite this, solar PV installation has grown fast – fastest in Germany, hardly the sunniest nation on the planet.

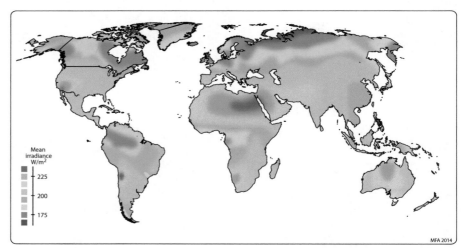

FIGURE 12.1
Map of the annual average solar irradiance.

[1]Jäger-Waldau, A., 2013. PV Status Report 2013, Report EUR 26118 EN, European Commission Joint Research Centre, Ispra, Italy, ISBN:978-92-79-32718-6. http://iet.jrc.ec.europa.eu/remea/sites/remea/files/pv_status_report_2013.pdf

Historically solar power has grown at 40% per year for the last 18 years. The efficiency of commercial solar panels, at present about 20%, is increasing and the price is falling. Their useful life is 25 years. Panels are rated by the peak power (subscript "p") they can produce under ideal conditions – thus a panel rated at $1\,kW_p$ will deliver $1\,kW$ under full sun at normal incidence. But the sun shines for only part of the day and then only on some days and, for part of that time, obliquely. In practice the power, averaged over time, is less than the peak by a fraction called the *capacity factor*, typically 0.05–0.15. The current (2014) installed PV capacity is $100\,GW_p$ world-wide, but because of the capacity factor the actual delivered power, averaged over time, is about one-tenth of this, meaning that solar PV at present contributes only 0.4% to total electricity consumption.

The potential contribution is much larger. Many national governments seek to realise this potential by offering subsidies and feed-in tariffs (FITs). Could solar PV provide the 20% to which they are committed?

Resources in the Appendix. Table A10 lists the energy mix in electrical power generation in selected nations. Table A12 gives the resource intensity of electrical power generation, including various forms of solar power. Table A13 characterises energy storage systems. Table A15 tabulates energy conversion efficiencies of fossil fuels to electricity and electrical energy to other forms of energy. Table A6 presents the abundance in the earth's crust and the world production of selected elements including some of those used in solar PV.

12.2   PRIME OBJECTIVE AND SCALE

Step 1

The government commitments define an objective, size and time scale: a 20% reduction in carbon emission from electric power generation by 2020. Is it achievable by solar PV alone? Is it a sustainable development? Those are the questions explored here.

One refinement. Solar power is intermittent. When it contributes only a small fraction of total power, the intermittency can be tolerated, but to rely on solar or wind for more than 20% of the total requires grid-scale energy storage to cope with the fluctuations. Economic storage on this scale is work-in-progress, not yet available, and it will require large-scale investment. Rather than speculate on this aspect, we will assume that PV deployment continues to grow at 40% per year until the 20% target is reached, at which point it stops.

12.3 STAKEHOLDERS AND THEIR CONCERNS

Contemporary news headlines suggest a range of stakeholders.

- *"Solar Power Begins to Shine as Environmental Benefits Pay Off".* Carol Olson of the Energy Research Center of the Netherlands notes that rising electricity prices make solar power increasingly attractive in Europe. (New York Times, 11 November 2013)
- *"Can solar panels really beat your pension?"* Energy minister Greg Barker, argues that putting money into solar panels on the roof of your house could deliver a better return than in a private pension. (The Daily Telegraph, London, 6 February 2014)
- *"End to solar farm blight as subsidy scheme is scrapped."* Green energy subsidies will be shut to large solar farms as ministers attempt to curb blight to countryside. (The Daily Telegraph, 13 May 2014)
- *"Edison signs deals for seven new solar power plants."* Energy provider Southern California Edison signs contracts for the construction and operation of seven solar power plants, one of them the largest single solar photovoltaic installation in the U.S. (Los Angeles Times, 11 January 2011)
- *"Solar stocks surge after Warren Buffett buys Calif. plants."* Investor interest in solar stocks boosts share price boom. (Los Angeles Times, 3 January 2013)

These quotes suggest the key stakeholders (Figure 12.2).

- Government, who see solar PV as a contribution to meeting renewable energy commitments and have the power, through subsidies and feed-in tariffs, to make it happen.
- Makers of PV systems, who see a growth market but a highly competitive one with diminishing profit margins.
- Retailers and distributors who see a market that is very sensitive to provision or withdrawal of subsidies.
- Investors, who see a rapidly expanding industry but one in which many smaller manufacturers and distributors fail.
- Energy providers, who see opportunities in renewable energy that, ultimately, will require installation of energy storage to smooth intermittent generation.

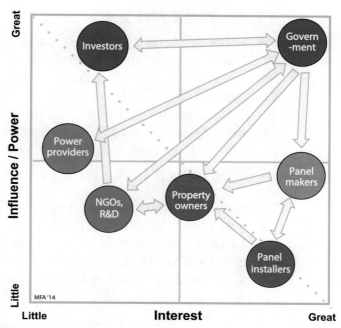

FIGURE 12.2
Stakeholders and influence.

- Property owners, attracted by feed-in tariffs and the sense of contributing to reduced carbon emission.
- NGOs that promote clean energy and campaign to halt climate change.
- R & D centers that see stimulation and funding for PV research.
- General public, concerned about visual intrusion of large solar farms.

SUMMARY OF THE SIGNIFICANT CONCERNS

- Cost – PV power is more expensive than conventional generation.
- "Solar panel blight" – solar PV installations are seen by some as visually offensive.
- The lack of transparency and continuity of subsidies and FITs, at least in the near term.
- Intermittency of generation, acceptable while scale is small but a problem if total contribution exceeds 20%.

These concerns flag the targets for the fact-finding stage that follows.

12.4 FACT-FINDING

Can solar PV make a significant contribution to carbon reduction by 2020? To answer that we must examine how quickly the carbon footprint of the solar cells themselves is offset by the lower emissions associated with the power they produce. This, the economics and societal acceptance appear to be barriers that have to be overcome if large-scale solar PV is to be sustainable. Figure 12.3 summarizes the issues needing research (Table 12.1).

$$r \times 20{,}000 \times P - C_f \times 24 \times 365 \times 8 \times P = (20{,}000r - 7008)\,P$$

(a) Materials.[2] Over 80% of present-day PV panels are based on high-purity silicon, consuming roughly 5.6 kg per kW_p. The anticipated world production of PV panels in 2013 was 35 GW_p, consuming some 196,000 tonnes of refined silicon, about 2.5% of the annual world production of raw silicon. The table shows where silicon is

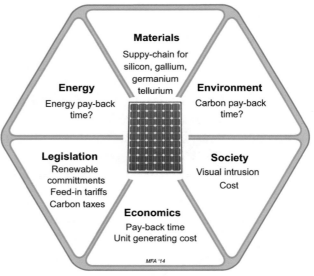

FIGURE 12.3
A fact-finding summary.

[2]Tritt, T.M., Bottner, H., Chen, L., 2008. Thermoelectrics: direct solar thermal energy conversion. MRS Bull 33 (4). Keoleian, G.A., Lewis, G.McD., 1997. Application of life cycle energy analysis to photovoltaic module design. In: Progress in Photovoltaics: Research and applications, vol. 5. Philipsen, G.J.M., Alsema E.A., 1995. Environmental life-cycle assessment of multicrystalline silicon solar cell modules. Netherlands Agency for Energy and the Environment (NOVEM), Report no. 95057.

Table 12.1 World Production of Silicon	
Silicon-Producing Nation	**1000 tonnes/year 2011**
China	5400
Russia	670
The United States	350
Norway	320
Brazil	230
France	140
South Africa	130
Other countries	543
World	**7783**

Minerals.usgs.gov/minerals/pubs/commodity

produced. The only disturbing feature is the degree of supply-chain concentration: 70% of world production from a single nation.

Current PV panels use two elements that appear on "critical" lists. One is silver. The world production of silver is 23,000 tonnes per year. The production of $10\,GW_p$ of PV panels per year (corresponding to 40% growth) requires about 15% of world production. The other critical element is indium, used as a transparent conductor. Its world production is 640 tonnes per year. The same panel production requires 200 tonnes, 31% of world production.

There are alternatives to silicon, notably gallium arsenide and cadmium telluride. All four elements are toxic and so require special handling and disposal.

(b) Energy. The manufacture of PV systems (panels plus inverter) requires energy and has a carbon footprint. Figure 12.4 shows how both have fallen as the efficiency of manufacture has improved.[3]

How long will it take to reach 20% of total electricity generation, at which point grid-scale energy storage becomes necessary? With a current installed PV capacity of $100\,GW_p$ and a capacity factor of 0.1, the current effective PV contribution is 10 GW. Global electricity demand is 2,400 GW, so 20% of this is 480 GW. With a growth rate r of 40% per year, the time to reach 480 GW is

---

[3]Kawajiri, K., Gutowski, T.G., Gershwin, S.B., 2014. Net CO_2 emissions from global PV development. Energy Environ. Sci.

$$t = \frac{100}{r} \left(\ln \frac{480}{10} \right) = 9.6 \text{ years}$$

We conclude that, if the current growth rate is maintained it is possible to reach the target of 20% PV generation by 2023.

But what about energy? Will installing PV systems at this rate reduce hydrocarbon consumption? Energy is invested in making the system at the start of its life but the energy it generates is spread over its operating life. If the rate of installation is too fast, the energy investment increases faster than the panels can compensate. The energy return on investment is then negative until construction stops, when it starts to recover. At a current installed capacity of P kW$_p$ and a growth rate r of 40% per year, and an embodied energy (from Figure 12.4) of 20,000MJ per kW$_p$, the construction of the panels requires

$$r \times 20{,}000 \times P = 8000 \, P \text{ MJ/year} \tag{12.1}$$

Meanwhile, the already installed panels, operating with a capacity factor $C_f = 0.1$, generate, electrical power that would, without the panels, have been drawn from the grid. For a valid comparison we must convert all energies to units of primary energy. Globally, 1 kW.hr of grid electricity requires about 8 MJ of primary (fossil fuel) energy. Thus the saving of primary energy made possible by the currently installed panels is

$$- C_f \times 24 \times 365 \times 8 \times P = - 7008 \, P \text{ MJ/year} \tag{12.2}$$

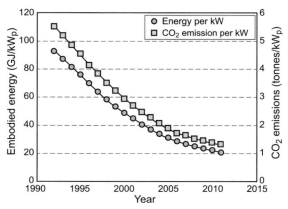

FIGURE 12.4
Embodied energy and carbon footprint of solar PV per kW of nominal generating capacity.

(the minus sign means a saving). Thus throughout the growth period, the panels are absorbing more energy per year than they are producing. If construction stops when the 20% target is reached in or near 2023, the energy deficit begins to be paid back following the pattern sketched in Figure 12.5 but there is still no saving of hydrocarbons until the break-even point is reached. There is a conflict between rapid PV deployment and environmental burden during growth. A slower rate of development can offer a steady decrease.

Exercise. What is the critical growth rate below which deploying solar PV systems results in a steady saving of energy?

Answer. Summing Eqns (12.1) and (12.2), with a growth rate r gives the energy balance:

$$r \times 20,000 \times P - C_f \times 24 \times 365 \times 8 \times P = (20,000r - 7008)P$$

The critical growth rate for zero energy change is

$$r = \frac{7008}{20,000} = 35\% \text{ per year.}$$

(c) The Environment. As Figure 12.4 shows, the carbon footprint of manufacture of solar PV systems is considerable – about 1200 kg per kW_p of generating capacity. The power it generates is carbon free but it is distributed over the life of the system. As with energy, a growth rate of 40% per year results in a net increase, not a saving, of carbon emissions. This deficit continues to grow until construction stops, following a pattern like that of Figure 12.5.

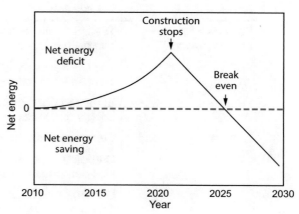

FIGURE 12.5
Schematic of energy demand from conventional power over time when growth is too fast.

(d) Legislation. Governments seek to drive the transition to low-carbon energy using both carrots and sticks. The carrots – subsidies and FITs – reward the installation of low-carbon systems. The sticks – carbon taxes and enforced carbon trading – penalize activities that are carbon intensive. Table 12.2 gives examples of current legislation.

(e) Economics. Does installing solar PV on your house make economic sense? Today (2014) the installed price of grid-connected solar PV is €1600 ($2300) per kW_p. At a capacity factor of 0.1, a 1 kW_p system will deliver 880 kWh/year. The FIT – what you are paid to feed electricity into the grid – has fallen from a high of €0.5/kW.hr in 2011 to an average of €0.12/kW.hr across much of the world in 2014 as system prices have fallen. Thus the 1 kW_p system generates a FIT income of €105/year. At this rate, how long will it take the system to break even in financial terms?

We explored the time-value of money in Chapter 4. The pay-back time, n^* years, on a capital investment P, repaid at the rate A/year at an interest (discount) rate i that we take to be 5% per year, is given by Eqn (4.5). Inserting the data above for a 1 kW_p system gives

$$n^* = \frac{-\ln\left(1 - \frac{iP}{A}\right)}{\ln\,(1+i)} = 29 \text{ years}$$

Table 12.2	Legislation Bearing on Solar PV Installation
Sector	**Legislation**
Subsidies for low-carbon power	The US Business Investment Tax Credit for renewable energy investment
	The US Recovery Act 1603 Program for business investment in renewables
	European Feed-In Tariffs: subsidies for renewable energy generation
Carbon taxation and trading	European carbon tax: a pollution tax on fossil-fuel use
	Carbon trading: an attempt to use market forces to limit carbon emissions
	Carbon off-setting: sale of credits in activities that limit or sequester carbon

This is more than the 25 year design life of the system.

Domestic electricity in Europe costs about €0.15/kW.hr, in the US it is about $0.17/kW.hr. If you use all the solar PV power that you generate yourself, you save that much per kW.hr in addition to the FIT. Then you do much better – the pay-back time falls to 8.4 years.

(f) Society. As one of the news-quotes indicated, solar PV is seen by some as a good investment. Government schemes, promoted by generous tax-breaks and FITs, have stimulated the rapid growth of domestic solar PV, widely accepted by the public. Larger-scale installations have grown even faster. Opposition to these is now emerging on the grounds of visual intrusion.

SUMMARY OF SIGNIFICANT FACTS

- The efficiency of solar PV systems is rising and the cost is falling.
- Silicon-based solar cells dominate the PV market. The supply chain for silicon shows significant concentration, with over 70% produced by a single nation.
- The installation of solar PV is at present growing so rapidly (40% per year) that the energy per year produced by existing installations is less than the energy invested per year in new ones.
- The present build-rate actually increases, rather than reduces carbon emissions. Only when construction of new installation stops will solar PV contribute to carbon reduction.
- The pay-back time for small-scale domestic solar PV is, with present subsidies, long (of order 20 years) if the power is simply sold to the grid. If, instead, the producer uses the power, the pay-back time falls to about 8 years.
- Several things can increase the attraction of solar PV: higher grid electricity prices, lower installation costs and greater PV panel efficiency.

12.5   SYNTHESIS WITH THE THREE CAPITALS

What impact do the facts have on the three capitals?

Natural capital. The potential of solar PV to provide low-carbon power is the largest among all of available renewable sources.[4] It is clean,

---

[4]European Photovoltaic Industry Association (EPIA), 2014. Global market outlook for photovoltaics 2014–2018. http://www.epia.org/news/publications/.

silent and reliable, with a 25 year useful life. But if expansion is rapid an energy deficit and a net increase in carbon emission arises. Compensation for both follows when the rate of expansion slows or stops.

Manufactured and Financial capital. Solar PV systems use well-established technology to provide reliable power with guaranteed output for at least 25 years. Subsidies and FITs, used by most national governments to stimulate growth in renewable energy, caused very rapid expansion. These tariffs have now been reduced, meaning that the financial pay-back time is at present long. But the efficiency of PV systems has increased, their price has decreased and the cost of grid-electricity has risen steadily over the last 20 years. If these trends continue, the price of PV and conventional utility electricity will approach parity.

At present, solar PV contributes about 0.4% of total world electricity generation. The current rate of growth, if maintained until 2023, would expand this to 20%. At this point further expansion would require investment in grid-scale storage to smooth the fluctuations in generating capacity between night and day and with changes of cloud cover. The cost of this additional resource is not, at present, known but is expected to be large.

Human and Social capital. All renewable energy systems require a large area if they are to produce sufficient power to make a real contribution to national needs. The visual impact of large solar farms is seen, by some, as undesirable. Domestic solar PV is less obtrusive and allows direct grid-connection, eliminating the need for additional distribution networks.

There is a personal sense of assurance that comes with the ability to provide, even in a small way, for one's own energy needs. Solar radiation is stable, and once the solar PV is installed, the price of the electricity it provides is also stable, independent of fluctuating energy prices for 25 years. To many, this is sufficient reason to install roof-top panels (Table 12.3).

12.6   REFLECTION ON ALTERNATIVES

Short term. Renewable energy on a significant scale requires investment not just of capital but of energy, most of it from conventional (fossil-fuel based) sources. The analysis of the case study highlights

Table 12.3 — Synthesis: The Impact of Solar PV Installation on the Three Capitals

	Human and social capital – People *Health? Wellbeing? Convenience? Culture?* *Tradition? Associations? Perceptions?* *Contributes to equality? Morality?*	Natural capital – Planet *Can prime objective be met?* *Are stakeholder concerns addressed?* *Are there unwanted consequences*	Manufactured capital – Prosperity *Cost – Benefit? (Cost facts vs. Eco facts)* *Legitimacy? (Conformity with law)*
Materials		(–) Some PV panels use critical elements for their manufacture: silver, indium, gallium, tellurium	
Energy	(+) There is an element of personal satisfaction in meeting energy needs locally	(+) Solar PV has great potential as a long-term source of renewable energy (–) Rapid expansion can require more energy than existing PV installation can , at the time, produce	(–) Solar PV electricity is at present more expensive than that from national grids (+) Solar PV seen as a cushion against fluctuating energy prices and uncertain supply (–) Expanding PV generation beyond 20% of national need will require grid-scale energy storage
Environment	(+) Solar PV widely accepted as one path towards renewable energy.	(+/–) Solar PV power is carbon free but manufacture of the hardware is not. Rapid expansion can cause temporary increase in carbon emissions	
Legislation			(+) Subsidies and feed-in tariffs (FITs) have stimulated very rapid expansion of solar PV installation
Economics	(+/–) Solar PV is promoted as a good investment: "better than your pension". It is not clear that this is true.		(–) The reduction of the FIT means the pay-back time for solar PV is at present long (+) Cost parity of PV and commercial utility electricity is on the horizon
Society	(–) Solar farms require a large area and are found objectionable by some, on visual grounds		(+) The solar PV industry creates high technology employment
Synthesis (the most telling facts)	(+) Although there are objectors, solar PV attracts less opposition than other types of renewable energy generation.	(+/–) The environmental picture is one of short-term loss for longer-term gain.	(+) The financial picture, like the environmental one, is that of short-term loss for long-term gain

the difficulty: if the transition is to be made in a time-scale measured in decades, the energy required to make and install the systems will largely cancel the contribution the renewables make to supply. Worse, the associated carbon emissions may cancel or even exceed the reduction that renewable sources, once established, can offer. It is a transition problem: once the renewable systems are in place and further installation comes to an end, the expected gains appear. The challenge, then, is to smooth the transition.[5] This is best done by

- making PV panels in countries with low CO_2/kW.hr of energy (Norway, Iceland, France);
- deploying PV panels where PV potential is high, that is, in sunny locations (Figure 12.1); and
- deploying PV panels where electricity mix has high CO_2/kW.hr, thus using the low-carbon PV power to greatest effect.

Longer term. Once a stable number of solar PV facilities are in place, the difficulty largely disappears. With a 25 year life, the requirement for new (replacement) systems falls to just 4% per year, one-tenth of the current build-rate. Further, with a known fraction of panels coming out of service each year, the establishment of a recycling or re-engineering program becomes practical.

The conclusion, then, is that the prime objective – that of a 20% reduction in carbon emission from electric power generation by 2020 – is not achievable by relying on solar PV. Indeed, at the present build-rate, it will increase carbon emissions rather than reduce them. It is a mistake, however, to interpret this as an argument against solar PV power. It is clean, reliable and free of the price fluctuations and potential supply-constraints of fossil-fuel based power. It is, as already said, a question of short-term loss in order to achieve long-term gains.

12.7 SUGGESTED PROJECTS

These projects draw on some of the information assembled in this case study, but the stakeholders, the operating requirements and expectations of users may differ significantly.

P12.1 Conduct an analysis like that of this case study on a national scale for the country in which you live. What is the solar energy flux (Figure 12.1)? What is the energy mix in electric

---

[5]Ibid.

power generation? What fraction is from solar PV? What is the build-rate of new solar PV? Does it exceed the critical rate for net energy gain and carbon reduction? Who are the stakeholders? Have they expressed their views in the press, on television or in other ways? What, in your view, is the potential for solar PV in your country?

P12.2 The nation with the fastest growing solar PV installation is Germany, with a current growth rate of 61% per year. Compare the situation in Germany with that in your own country. How do the levels of solar radiation compare in the two countries? (Figure 12.1 will help.) How do people's expectations and priorities for electrical power differ in the two countries? What are the differences in legislation and the consistency or otherwise of support for renewables via FITs? How is it that the build-rate in Germany is larger than that in your own country?

Further Reading

Ashby MF: *Materials and the Environment*, Table 6.8, 2nd ed., Oxford, 2013, Butterworth Heinemann.

European Photovoltaic Industry Association (EPIA): *Global Market Outlook for Photovoltaics 2014–2018*. http://www.epia.org/news/publications/, 2014.

Gutowski TG, Gereshwin SB, Bounassisi T: *Proc IEEE Int Symp Sustainable Systems and Technol*, 2010. Washington DC.

Jäger-Waldau A: *PV Status Report 2013*, Report EUR 26118 EN Ispra, Italy, 2013, European Commission Joint Research Centre, ISBN: 978-92-79-32718-6. http://iet.jrc.ec.europa.eu/remea/sites/remea/files/pv_status_report_2013.pdf.

Kawajiri K, Gutowski TG, Gershwin SB: Net CO_2 emissions from global PV development, *Energy Environ Sci*. 2014.

MacKay DJC: *Sustainable Energy – without the Hot Air*, Chapter 9. Cambridge, 2008, UIT, ISBN: 978-0-9544529-3-3 (Clear thinking and common sense applied to energy generation and use.).

US Office of Energy Efficiency and Renewable Energy. http://energy.gov/eere/energybasics/energy-basics, 2014.

CHAPTER 13

Bamboo for Sustainable Flooring

Chapter Outline

13.1 Introduction and Background Information 198

13.2 Prime Objective and Scale 200

13.3 Stakeholders and Their Concerns 200

13.4 Fact-Finding 202

13.5 Synthesis with the Three Capitals 207

13.6 Reflection on Alternatives 208

13.7 Suggestions for Related Projects 210

Further Reading 210

Materials and Sustainable Development. http://dx.doi.org/10.1016/B978-0-08-100176-9.00013-X
Copyright © 2016 Elsevier Ltd. All rights reserved.

13.1 INTRODUCTION AND BACKGROUND INFORMATION[1]

Bamboo is a grass, not a tree. It grows in an equatorial belt that circles the earth (Figure 13.1). Bamboo is exceptionally fast-growing, allowing it to be harvested for construction when 4–7 years old. The world production of bamboo is roughly 1.4 billion tonnes per year. International trade in bamboo in 2005 was valued at $2.5 billion and is projected to rise to $20 billion by 2020.[1] About 2.5 billion people in the world depend on it economically. These are big numbers – few other materials are used in such quantities.

China[2] and India are the largest producers and users of bamboo, but many other nations grow it commercially. It is used for scaffolding and construction as bamboo poles in countries in which it is indigenous. In Europe and North America it is available as laminated bamboo board (Figure 13.2), a higher value-added product that is better adapted for construction purposes in those countries.

Proponents of bamboo claim that its properties are comparable to those of hardwoods. They argue for its greater use in construction

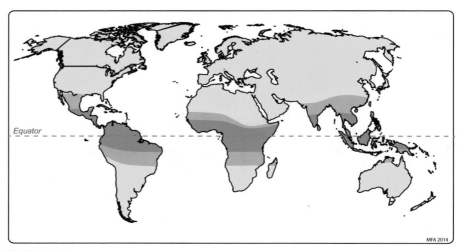

FIGURE 13.1
The global bamboo belt.

[1]FAO (2007) "World Bamboo Resources – A Thematic Study Prepared in the Framework of the Global Forest Resources Assessment 2005". US Food and Agriculture Organization. ftp://ftp.fao.org/docrep/fao/010/a1243e/a1243e00.pdf
[2]http://www2.bioversityinternational.org/publications/Web_version/572/ch10.htm

FIGURE 13.2
A bamboo-board panel.

in Europe because it is sustainable and because it is sourced from some of the world's least prosperous nations, where it creates employment. Consider, then, a possible scenario based on this information: that of doubling the use of bamboo board for flooring and walls in domestic and office buildings in Europe.

13.1.1 Background Information

Bamboo board is made from compressed bamboo. Large bamboo poles are cut into strips and stacked with the grain at right-angles. The stack is impregnated with a low VOC resin and pressed and cured. The resulting board is milled and sanded to give finished panels. Property data for the properties of bamboo board are limited. There are three grades – Super, First and Second class – spanning the ranges shown in Table 13.1. Bamboo chip board has less good properties.

Table 13.1	Properties of Bamboo Board	
Property[a]	**Value**	**Units**
Price	1.9–2.2	$/kg
Density	600–872	kg/m^2
Flexural modulus	7.2–10	GPa
Flexural strength	45–95	MPa
Hardness	57–59	MPa
Thermal conductivity	0.16–0.2	W/m.K

[a]Sulastiningsih Nurwati IM "Physical and mechanical properties of laminated bamboo board" Journal of Tropical Forest Science 21(3): 246–251 (2009)

Resources in the Appendix. Table A8 gives production and price data for diverse materials including bamboo. Table A19 lists embodied energy, carbon footprint and water usage of the same materials. Tables A20 and A21 give economic and governance-related information for nations, including the principal bamboo producers.

13.2   PRIME OBJECTIVE AND SCALE

The scenario suggested here is that of doubling the use of bamboo board in Europe. For an estimate of the time frame we need what might be called logarithmic logic. A time of 1 year for a major project requiring infrastructure is impossibly short. Hundred years exceeds the lifetime of the stakeholders. Split the difference (logarithmically): the obvious answer is 10 years. If this sounds cavalier, think for a moment of typical government and international horizons of thought. Most recent commitments on climate change agreed prior to 2014 set 2020 as their target. But now that we are half way there, a target of 2030 starts to appear.

13.3   STAKEHOLDERS AND THEIR CONCERNS

Here are contemporary news quotes about bamboo

- *"Why Farm Bamboo? Farming bamboo as an income producing crop. It is a perennial. You do not have to replant it each year. It is evergreen and removes carbon dioxide from the air even in winter. Bamboo reduces erosion and protects fields from wind"* (Bamboo farming USA[3], June 2013).
- *"What is World Bamboo? The World Bamboo Organization is a diverse group consisting of individual people, commercial businesses, non-profit associations, institutions, and allied trade corporations that all share a common interest, BAMBOO"* (World Bamboo[4], 6 May 2014).
- *"Focus Northeast: Freeing the bamboo trade can galvanize the region. Northeast India is a region of abundant natural resources, with 45% of global bamboo reserves, yet the lack of development here is palpable. Freeing the bamboo trade can galvanize the rural economy"* (The Times of India, 21 April 2014).

[3]www.bamboofarmingusa.com/
[4]http://worldbamboo.net/

- *"Bamboo is called the poor man's wood because it is used in all aspects of rural life"* (Bamboo Select[5], 2011).
- *"* Is Bamboo Flooring a Good Idea or Just a Fad?" Steve Appolloni, a designer in Albuquerque, N.M., likes it for its durability, cost and eco-friendly qualities (New York Times, 17 April 2008).
- *"It looks and feels like timber flooring, but you could be walking on bamboo. Bamboo offers an eco-friendly alternative to hardwood for a fraction of the price"* (The Sunday Telegraph, 29 April 2014).
- *"Bamboo in Construction: Is the Grass Always Greener? The lack of codes and standards has kept architects and designers away from bamboo, according to a 1997 INBAR (International Network for Bamboo and Rattan) newsletter"* (Environmental Building News[6], 1 March 2006).
- *"Bamboo: a new approach to carbon credits. Bamboo is now officially recognized as a carbon offset and as a tool for climate change mitigation"* (International Network for Bamboo and Rattan, INBAR, 27 November 2012).

These quotes suggest the key stakeholders (Figure 13.3).

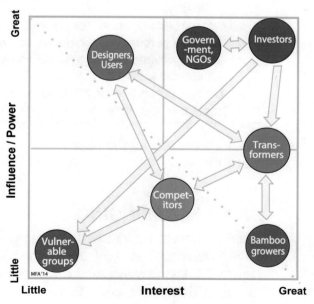

FIGURE 13.3
Stakeholders and influence.

[5]http://eng.bambooselect.com/Bambooworld.html
[6]http://www2.buildinggreen.com/article/bamboo-construction-grass-always-greener-0

- Bamboo growers – farmers and communes in agricultural communities for whom bamboo culture is a profitable business.
- Transformers – makers of bamboo-based products who seek greater acceptance and a secure supply chain.
- Investors – banks and other capital sources, wishing to see growth and profitability
- Supporters – local and national governments and NGOs, who seek job creation and export potential
- Users – architects, designers and property owners, who are influenced by aesthetics, associations and the negative perception of bamboo.
- Competitors – makers of conventional flooring, who might wish to portray bamboo as an inferior material.
- Vulnerable groups – people displaced by change of land use to bamboo cultivation or from industries dependent on the displaced crop.

SUMMARY OF THE SIGNIFICANT CONCERNS

- Restrictive practices and lack of investment constrain bamboo cultivation.
- Lack of building codes for bamboo deters designers and architects.
- Negative perception of bamboo as "poor man's wood".
- Are the environmental and economic benefits of bamboo culture sufficiently valued?
- Are the properties of bamboo board "as good as hardwood at a fraction of the price"?

These concerns flag the targets for research in the fact-finding stage that follows.

13.4   FACT-FINDING

What information is needed to support the claim that bamboo board performs as well as hardwood but is more sustainable? Figure 13.4 summarizes the issues needing research. No judgements at this point, only facts.

(a) *Material.* Are the properties of bamboo board really "comparable with hardwoods"? Its bending stiffness and bending strength, from Table 13.1, are compared in Figure 13.5 with those of beech plywood and solid oak, both commonly used for flooring. The figures demonstrate that the claims are justified.

European hardwoods like oak are widely available. Is bamboo as accessible as they are? The main bamboo-producing nations are

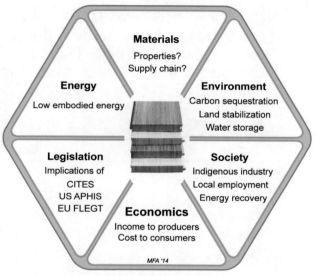

FIGURE 13.4
A fact-finding summary.

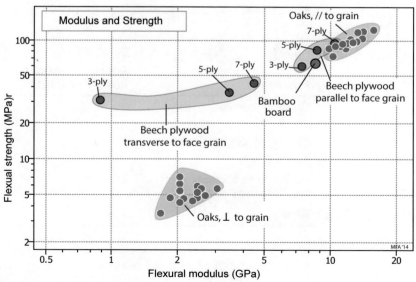

FIGURE 13.5
A comparison of the properties of bamboo board and woods.

listed in Table 13.2. At first sight there appears to be supply-chain concentration: China is the biggest exporter of bamboo and bamboo products, with 80% of world trade. But, as the quote from the Times of India (above) makes clear, the unexploited reserves of bamboo elsewhere are very large.

(b) *Energy*. A detailed LCA[7] is available for bamboo grown and processed into board in China then shipped to Europe. The results are summarized in Table 13.3. The energy and emissions of cultivation are tiny. Those of processing to board are considerable. Shipping to Europe is about 20% of the total.

Table 13.2	Bamboo-Producing Nations and Stock
Nation	**Bamboo Stock in 2005 (Tonnes)[a]**
China	1,230,000,000
India	14,615,000
Myanmar	9,830,000
Nigeria	7,320,000
Sri Lanka	1,500,000
Bangladesh	1,100,000
Indonesia	220,000

[a]FAO (2007) "World Bamboo Resources" ftp://ftp.fao.org/docrep/fao/010/a1243e/a1243e00.pdf.

Table 13.3	Energy and Carbon Footprint of Bamboo Board		
Process	**Approx. Embodied Energy (MJ/kg)**	**CO_{2eq} (kg/kg)**	**% of Total**
Growing, harvesting, local transport	0.3	0.04	3
Slicing, drying, bonding, finishing	13	1.13	80
Transport to Europe (19,000 km by sea)	2.1	0.24	17
Total	15.4	1.41	100

[7]van der Lugt, P. Vogtländer, J.G., van der Vegte, J.H. and Brezet, J.C.(2012) "Life cycle assessment and carbon sequestration; the environmental impact of industrial bamboo products" IXth World Bamboo Congress, Belgium, pp73–85. http://www.worldbamboo.net/wbcix/presentation/ Van%20der%20Lugt, %20Pablo.pdf

Is bamboo board less energy intensive than other types of floor cover[8]? Figure 13.6 compares the embodied energy of flooring per functional unit, which we take to be 1m² of floor in Europe. Qualifications are needed here. Transport energy is about one-fifth of the total for bamboo flooring; most of the other floor materials are sourced in or nearer to Europe – their energies are "at source", so may be underestimates. If, at the end of life, the bamboo flooring is combusted with energy recovery, there is a credit that cannot be claimed for most of the others. The somewhat fuzzy picture that emerges is that bamboo board has an embodied energy that, relative to other floorings, is low.

(c) *The Environment.* Assessing carbon emission for natural materials involves three interesting distinctions.

- *Biogenetic CO_2* is absorbed from the atmosphere when bamboo grows and is released when it decays or is burnt. The net result is zero – it is carbon-neutral.
- *Processing CO_2* is that released during cultivation, harvesting, drying and board-making. It is a debit.

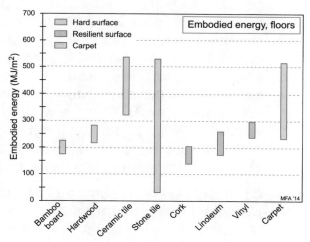

FIGURE 13.6
The embodied energies of flooring materials.

[8]Holtzhausen, H. J. (2007). Embodied energy and its impact on architectural decisions. 3rd international Conference on Sustainable Development Planning, 25–27 April 2007. 102, pp. 377–385. Algrave, Portugal: WIT Transactions on Ecology and the Environment.

■ *Recovery CO_2* is the credit accessible by incinerating bamboo at end of useful life with energy recovery for electric power generation. The credit arises because the power generated in this way is no longer drawn from a conventional fossil-fuel power station. That released by burning has already been allowed for under "Biogenetic CO_2".

The total carbon footprint of bamboo board, like its embodied energy, is low. But that is not the only environmental service offered by bamboo. The Kyoto Protocol of 1997 recognized forests as carbon sinks. Vigorous growth makes bamboo a particularly attractive plant for carbon sequestration. The below-ground root system stabilizes sub-soil and is said to store water. The decaying foliage provides biomass.

(d) *Legislation.* If bamboo board is made in bamboo-rich nations and exported to the US and Europe (the two largest markets) the trade must conform to international and local controls. Many nations restrict the imports of plants, animals and products derived from them to safeguard their own agriculture and natural habitat. Table 13.4 lists examples.

(e) *Economics.* The estimated annual production of bamboo flooring in China[9] in 2005 was 17.5 million m^2. In the EU, consumption of bamboo flooring has increased from 0.67 million m^2 in 2003 to 0.9 million m^2 in 2005 and is expected to continue growing. Bamboo board costs \$5 to \$6.50 per ft^2 (about \$50 per m^2) in the US and about €30 per m^2 in Europe, a little less than oak flooring (about €30 or \$60 per m^2). Thus the world trade in bamboo board has a value of at least \$500 million per year, though, of course, only a part of this reaches the supplying nations. The claim that bamboo flooring costs "a fraction of the price of hardwoods" is not supported.

(f) *Society.* Bamboo culture and processing creates employment in planting, primary and secondary processing, construction and the manufacture of higher-end products like flooring, furniture and fabrics[10]. Many of the world's least prosperous nations lie within the bamboo belt, allowing the development of bamboo-based

---

[9]http://www.fao.org/docrep/010/a1243e/a1243e00.htm and http://www.ecoplanetbamboo.net/files/bamboo_worldwide.pdf

[10]http://www.ecoplanetbamboo.net/files/bamboo_worldwide.pdf

Table 13.4	Legislation Bearing Imports and Exports of Plants and Animals
Sector	**Legislation**
International	*CITES Convention on International Trade in Endangered Species* gives various degrees of protection to 35,000 species of plants and animals
US legislation	*APHIS-PPQ Plant Protection Quarantine* regulates imports of plants, including woods
	APHIS restrictions on imports of wood require that woods be heat or chemically treated before import
Europe	*The EU Plant Health Directive (2000)* restricts imports of plants, products and materials deemed to be harmful to agriculture and native species.
	The EU FLEGT Action Plan (2003) aims to halt imports of illegally logged woods. Further EU restrictions prohibit woods known to host harmful pests or diseases. Currently controlled species include all conifers, maple, sweet chestnut, oak, plane and poplar (aspen).

industries that support the local economy or, as the Times of India mentioned, could do so.

SUMMARY OF SIGNIFICANT FACTS

- Bamboo cultivation provides a range of environmental services – carbon sequestration, land stabilization, biomass.
- Bamboo cultivation provides employment and income in some of the world's least prosperous regions.
- The properties of bamboo board and the price per m^2 are comparable with those of hardwoods.
- Bamboo board is a relatively high-end product.
- Expanding bamboo board production is impeded by lack of will, investment and training.
- The image of bamboo as "poor man's wood" is perceived to disadvantage it.

13.5 SYNTHESIS WITH THE THREE CAPITALS

What impact do the facts have on the three capitals? Some impact on more than one capital; for simplicity, we assign the influence to the one most affected. As explained in Chapter 3, there is no single answer to this question; the weight given to the facts depends on values, culture, beliefs and ethics. Here we present one view. The main points are summarized in Table 13.5 with positive (+) and negative (−), positive influences color-coded.

Natural capital. If grown under the correct conditions, bamboo's green credentials are impressive. It prevents soil degradation and erosion, requires little water, and sequesters carbon. If used for high-end wood products, such carbon remains stored for long time periods. Increasing the harvesting of bamboo is possible by drawing on untapped natural bamboo forests.

Manufactured and Financial capital. If bamboo is so good, why is it not more widely exploited? The market for bamboo outside of China is in its infancy. A disconnect between agronomists, financiers and potential end users, aggravated by trade restrictions on plant products, has slowed commercialization.

Human and Social capital. Sustainable bamboo plantations provide direct employment for many rural, unskilled people in areas where opportunities for economic development are low, both in the plantations and in processing facilities. Carbon credits provide additional potential for poverty alleviation and economic diversification, but are frequently misused[11]. The negative perception of bamboo as a down-market material could be countered by greater emphasis on its environmental and aesthetic qualities.

13.6   REFLECTION ON ALTERNATIVES

Short term. The current picture is a positive one. There are more positive aspects than negatives for bamboo board as an alternative construction material and it is one that can be marketed as both green and aesthetically pleasing. The impediments to doubling the production of bamboo board over the next 10 years (the Prime Objective) appear to be trade barriers and the lack of capital investment.

Longer term. The most telling negatives are those that impact on manufactured capital. All could be resolved by action at

---

[11]Rogers, H. (2010) "Green Gone Wrong – How Our Economy is Undermining the Environmental Revolution", Scribner, New York.

Table 13.5	Synthesis: Impact of Increased Bamboo Board Manufacture on the Three Capitals		
	Human and social capital – People *Health? Wellbeing? Convenience? Culture? Tradition? Associations? Perceptions? Contributes to equality? Morality?*	**Natural capital – Planet** *Can prime objective be met? Are stakeholder concerns addressed? Are there unwanted consequences*	**Manufactured capital – Prosperity** *Cost – Benefit? (Cost facts vs. Eco facts) Legitimacy? (Conformity with law)*
Materials	(+) Bamboo cultivation employs unskilled labor and attracts carbon credits (−) Perception and associations of bamboo as low-end material an obstacle to development		(+) No obvious material resource constraints (−) Properties of bamboo board are poorly characterized
Energy		(+) Demand for energy and water are low	
Environment	(−) Lack of advice and training on crop management impedes production	(+) Offers several environmental services: renewable, carbon sequestration, soil stabilization.	
Legislation			(−) Lack of legal framework and widely accepted standards (−) US and EU regulation of imports of natural or illegally harvested materials an impediment.
Economics	(+) High value-added bamboo board generates income.		(+) Bamboo growth has a low economic overhead (+) Bamboo board is a high value-added product (−) Economic potential exists but capital investment is lacking
Society	(−) Need for training, skills, research not met. (+) Bamboo cultivation and board manufacture creates jobs	(−) Possible competition with other crops for land area	(−) Acceptability by consumers is questionable
Synthesis (the most telling facts)	Bamboo and bamboo products make positive contribution to human and social capital	Bamboo and bamboo board have impressive environmental credentials	A disconnect between agronomists, financiers and potential end users has resulted in the s low commercialization. Better characterization, codes and standards needed. Re-evaluation of aesthetic qualities needed.

government or state level to stimulate training, to enable capital investment and to develop new markets for bamboo products. Obstacles to international trade (import restrictions, burdensome reporting to establish legal harvesting, mandatory inspection etc.) could perhaps be reduced by negotiation at the government level.

13.7 SUGGESTIONS FOR RELATED PROJECTS

These projects draw on some of the information assembled in the Bamboo board case study, but the prime objective, the stakeholders, the producing nations and the economics are significantly different.

P13.1 Follow the method of this case study to explore the potential for expanding production of cork flooring. Is cork flooring more sustainable than other flooring materials? How much is produced and where? Is it, like bamboo, a staple of low income economies? What is its eco-character? Would doubling production be a sustainable development?

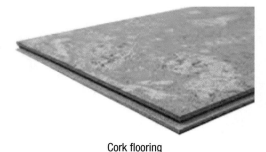

Cork flooring

P13.2 Mexico at present grows very little bamboo, although its climate would allow it. What is the potential for the development of a sustainable bamboo industry on a large scale in Mexico?

Further Reading

Woolley T: *Natural building – A guide to materials and techniques*, Ramsbury, UK, 2006, The Crowood Press Limited, ISBN: 1-861-26841-6 (A well-illustrated introduction to traditional building materials and their present day modifications.).

CHAPTER 14

The Vision: A Circular Materials Economy

Materials and Sustainable Development. http://dx.doi.org/10.1016/B978-0-08-100176-9.00014-1
Copyright © 2016 Elsevier Ltd. All rights reserved.

Chapter Outline
14.1 Introduction and Synopsis 212
14.2 The Ecological Metaphor 213
14.3 The Scale of the Vision 217
14.4 The Circular Materials Economy 219
14.5 Creating a Circular Materials Economy 222
14.6 Summary and Conclusions 234
14.7 Exercises 236
Further Reading 237

14.1 INTRODUCTION AND SYNOPSIS

We live at present with a largely linear materials economy. Our use of natural resources is characterized by the sequence *"take – make – use – dispose"* as materials progress from mine, through product, to landfill. Increasing population, rising affluence and the limited capacity for the planet to provide resources and absorb waste argue for a transition towards a more circular way of using materials. When products come to the end of their lives the materials they contain are still there. Repair, reuse and recycling (the three "Rs") can return these to active use, creating a technological cycle that, in some ways, parallels the carbon, nitrogen cycle and hydrological cycles of the biosphere. Repair, reuse and recycling are not new ideas; they have been used for centuries to recirculate materials and, in less-developed economies, they still are. But in developed nations they dwindled as the cost of materials fell and that of labor rose over time, making all three Rs uneconomic. So what is novel about the contemporary idea of a *circular materials economy*? Have not we been there before?

The answer perhaps is an attitude of mind. As McDonough and Braugart, two vocal proponents of circular thinking, point out waste is a symptom of design-failure. Governments have sought to reduce waste by take-back regulations, mandatory recycling targets and minimum service lives. They are a sort of chastisement: we are behaving badly; regulation carries the message that we must behave better. One reaction has been "efficiency" movements – eco-efficiency, material-efficiency, energy-efficiency – all admirable but all seeking to allow business as usual with reduced drain on natural resources and sense of guilt.

The "circularity" concept is a way thinking that looks not just for efficiencies but also for new ways of providing the functions we need that do not need regulation to control them. In the last decade momentum has gathered about this transition. The idea of deploying rather than consuming materials, of using them not once but many times, and of redesign to make this a reality has economic as well as environmental appeal. Governments now sign up to programs to foster circular economic ideas and mechanisms begin to appear to advance them. In this chapter we examine the background, the successes and the difficulties of implementing a circular materials economy.

14.2   THE ECOLOGICAL METAPHOR

Natural and industrial systems have some features in common and others that are strikingly different. Consider three.

The first: both the natural and the industrial systems transform resources, meaning materials and energy, to provide functionality. Nature does it through biological growth. Industry does it through manufacture. The plant kingdom captures energy from the sun, carbon dioxide from the atmosphere and minerals from the earth to create carbohydrates; the animal kingdom derives its energy and essential minerals from those of plants or from each other. The industrial system, by contrast, acquires most of its energy from fossil fuels and its raw materials from those that occur naturally in the earth's crust, in the oceans and in the natural world.

The second similarity: both systems generate waste, the natural system through the metabolism and death of organisms, the industrial system through the emissions of manufacture and through the finite life of the products it produces. The difference is that the waste of nature is recycled with 100% efficiency, drawing on renewable energy (sunlight) and natural decay processes to return it to the ecosystem. The waste of industry is recycled much less effectively, requires non-renewable energy (fossil fuels) to do so, and leaves a legacy of depleted resources and contaminated eco-systems.

The third similarity: both the natural and the industrial systems exist within a single global eco-sphere. This eco-sphere provides the raw materials and other primary resources, acts as a reservoir for waste and provides the essential environment for life, meaning fresh water, a breathable atmosphere, tolerable temperatures and protection from UV radiation. The natural system manages, for long periods, to live in balance with the eco-sphere. Our present industrial system, it appears, does not. Are there lessons to be learned about managing industrial systems from the balances that have evolved in nature? Can nature give guidance, or at least provide an ideal?

Of the 92 usable elements of the Periodic Table, four are of central importance for life: carbon, nitrogen, hydrogen and oxygen. They largely make up the carbohydrates, fats and proteins on which life depends. To allow a steady-state these elements must be constantly recycled around the eco-system. The most important of these circular paths are the *carbon cycle*, the *nitrogen cycle* and the *hydrological (water) cycle*. Subsystems have evolved that provide the links in the cycles and do so at rates that match, thus avoiding bottle-necks at which waste accumulates. Figure 14.1 is a sketch of one of these, the carbon cycle. Carbon dioxide in

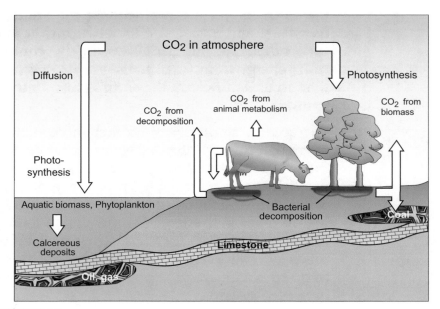

FIGURE 14.1
The carbon cycle in nature, showing some of the many subsystems that have evolved transform resources.

the atmosphere is captured by green plants and algae on land, and by phytoplankton and other members of the aquatic biomass in water. Fungal and bacterial action enables decomposition of plants and animals when they die, returning much of the carbon to the atmosphere but also sequestering some as carbon-rich deposits of peat, gas, oil, coal and, in the oceans, limestone, $CaCO_3$.

The elements important for man-made products, by contrast, are far more numerous. Today, as we saw in Chapter 1, they include most of the elements of the periodic table. Carbon is one. The products that use it (in the form of coal, oil or gas) eject most of it into the atmosphere as CO_2. As industrial activity has grown the rate of CO_2 emission has increased but the natural subsystems that recycle it have not evolved at a matching rate (Figure 14.2). When rates do not match, stuff piles up somewhere. Focusing on carbon again, this imbalance is evident in the steep rise of atmospheric carbon since 1850. The problem is more acute with other elements, the heavy metals for examples, for which no natural subsystems exist to provide recycling.

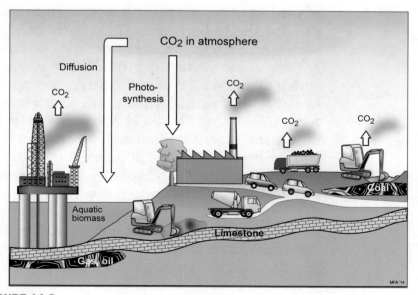

FIGURE 14.2
The additional burden placed on the carbon cycle by large-scale industrialisation.

Table 14.1 The Comparison of the Natural and the Industrial Eco-Systems	
The Natural Eco-System	**Present Day Technological Systems**
Uses few elements (mainly C, N, O and H)	Uses most of the periodic table
Multiply connected subsystems that use "waste" as a resource and self-adjust	Lacks subsystems that use "waste" as a resource and adapt automatically
Is non-linear – rich in feedback loops; resilient in times of stress	Is linear – transforms materials into products and waste; lacks resilience
Closed loop – no waste: each subsystem provides sustenance for others	Open loop – waste-accruing, destructive of sources on which it depends
Indicator of well-being: equilibrium	Indicator of well-being: growth

What do we learn? Table 14.1 summarizes the differences between the natural and industrial systems, the one, sustainable over long periods of time; the other, in its present form, not so. Of the many aspects of sustainability, two relate directly to materials. The first is the lack of appropriate subsystems to close many of the recycling paths, and second is that, where subsystems exist, there is an imbalance of rates. At base, however, there is a more fundamental difference: it has to do with metrics of well-being. That of nature is achieving *balance,* such that the system is in equilibrium. That of the industrial system is *growth.* Growth has become our measure for the well-being of businesses, nations and society as a whole. An economy that is growing is "healthy", one that is static is "sick". The present-day model for economic growth carries with it the need for ever-increasing consumption of materials and energy and of waste-creation. The characteristic that makes this system damaging for the environment is its insatiable appetite: the faster it ingests resources, the greater is the output of products and the better is its "health", even though this does not coincide with the health of the biosphere.[1] The comparison, then, suggests an ideal: an industrial system in which the consumption of materials and energy and the production of waste are minimized, and the discarded material from one product or process becomes the raw material for another.

---

[1]Frosch, cited by Guidice, F., La Rosa, G., Risitano, A., 2006. Product design for the environment – a life-cycle approach. Department of Industrial and Mechanical Engineering, University of Catania, Italy.

News-clip. "A new consuming philosophy: reuse, remake, refrain. Advocates of collective consumerism oppose waste and want you to share your goods or just buy less."The LA Times, 29 June, 2013.

News-clip. "Business leaders in Davos keen to mainstream the circular economy. Chief executives said biggest barrier to progress is ingrained mind-set of 'manufacture, use and dispose'." The Guardian, 30 April, 2014.

14.3   THE SCALE OF THE VISION

Figure 14.3 introduces differing scales of thinking about the relationship of technology and the environment. The horizontal axis describes the time scale, ranging from that of the life of a product to that of the span of a civilization. The vertical axis describes the spatial scale, again ranging from that of the product to that of society as a whole. It has four nested boxes, expanding outward in conceptual scale, each representing an approach to thinking about the environment.

The least ambitious of these – the smallest box – is that of *pollution control and prevention* (PC and P). This is intervention on the scale

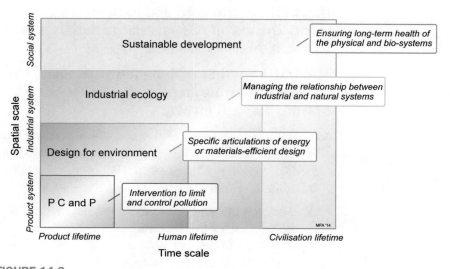

FIGURE 14.3

Approaches, differing in spatial and temporal scale of thinking, about the industrialisation and the natural ecosystem. Adapted from Coulter et al. (1995).

and life-time of a single product and is frequently a clean-up measure. Taking transport as an example, it is the addition of catalytic converters to cars, a step to mitigate an identified problem associated with an existing product or system.

The next box is *design for the environment*. Here the time and spatial scales include the entire design process; the strategy is to foresee and minimize the environmental impact of product families at the design stage, balancing them against the conflicting objectives of performance, reliability, quality and cost. Retaining the example of the car, it is to redesign the vehicle, giving emphasis to the objectives of minimizing emissions by reducing weight and adopting an alternative propulsion system – hybrid, perhaps, or electric.

The third box, that of *industrial ecology*, derives from the perception of human activities as part of the global eco-system. Here the idea is that the processes and balances that have evolved in nature might suggest ways to reconcile the imbalance between the industrial and the natural systems, an idea known as the ecological metaphor. The transport analog might be that of re-examining the concept of private transport, abandoning private ownership of vehicles and replacing it by a publicly owned system (driverless electric taxis, perhaps, charged from renewable-energy sources) that both fill the need and have minimal impact on the environment. The vision of a circular materials economy, described in a moment, fits into this box.

The last box, that of long-term *sustainable development*, requires thinking on a truly grand scale. Here is one (of many) formulation of what that will involve.

A set of operating principles for true sustainable development. Daly[2] (1990) and others[3] suggest the following operating principles for sustainable development

- The rate of use of renewable resources (air, water, biomass) must be no greater than the rate of regeneration.
- The rate of use of non-renewable resources (high quality minerals, fossil fuel) must be no greater than the rate at which renewable resources, used sustainably, can be substituted for them.

- The rate of emission of pollutants (gas, liquid and solid emissions) must be no greater than the rate at which they can be recycled, assimilated or degraded by the environment.
- People are not subject to conditions that undermine their ability to meet their own needs.

[2]Daly, H., 1990. Toward some operational principles of sustainable development. Ecol. Econ. 2, 1.
[3]The Natural Step, 2014. www.naturalstep.org.

Principles such as these require changes in the way we use materials that appear, today, to be unachievable. But they remain an ideal, something against which sustainability measures might be judged. For now we will stay with a more realistic vision.

14.4   THE CIRCULAR MATERIALS ECONOMY[4]

Before the industrial revolution, materials were expensive and labor was cheap. The number of materials in service was small and their high value relative to labour ensured that the products made from them were maintained, repaired and upgraded. Material efficiency was a normal practice.

Since the industrial revolution, large-scale mining and global trade have allowed material costs to decline[5] (look back at Figure 5.1). At the same time labor costs rose with the result that manufacturing methods (mass production, robotic assembly and the like) evolved to minimise the use of labor. The availability and low cost of materials encouraged industry to adopt an increasingly open-loop approach to material resources, transforming them into products that are used once then discarded; why conserve materials when they will be cheaper tomorrow than they are today? The flow of materials through the economy followed a linear path summarized as: *take – make – use – dispose* (Figure 14.4).

Some 80% of material usage still follows such a path but there is an increasing awareness that this cannot go on. The global population is increasing. Three billion consumers from the developing

[4]McDonough, W., Braungart, M., 2002. Cradle to cradle, remaking the way we make things. North Point Press, New York.
[5]World Bank, 2013. http://blogs.worldbank.org/prospects/category/tags/historical-commodity-prices.

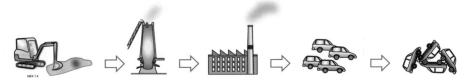

FIGURE 14.4
The linear materials economy: *take – make – use – dispose*.

world will enter the middle class by 2030,[6] with associated demand for products and services. The global consumption of materials, at present about 77 billion tonnes per year, is expected to rise to 100 billion by 2030.[7] The stress on the global eco-sphere caused by industrial development already gives cause for concern. And – a more immediate driver – the cost of materials is now increasing considerably faster than that of labor (Figure 5.1 again). An incentive is emerging to conserve materials rather than reject them. Increased material efficiency, reducing the resources per unit of manufacturing output, can make non-renewable resources last longer but their loss at the end of product life is a continuing drain. We examine ways to increase material efficiency in the next section, but first we explore the bigger picture: that of establishing a more circular materials economy.[8,9]

Materials in a circular economy are seen, not as a disposable commodity, but as a valued asset to be tracked and conserved for reuse, in rather the same way that financial capital is invested, recovered as revenues and re-invested. Figure 14.5 introduces the idea. Materials are produced and manufactured into products that enter service where they remain for their design life. In the circular model, disposal at end of life as landfill or waste is not an option. Instead the product is reused in a less demanding way, or reconditioned to give it a second lease of life, or dismantled into its component materials for recycling. All three of these options retain the materials of the product as *active stock* (the upper green box on the figure). It may be impractical to recycle perishable materials; those that are bio-degradable can be composted, returning them to the bio-sphere; those that are combustible

[6]McKinsey Quarterly, February 2014. Available from: www.ellenmacarthurfoundation.org.
[7]http://www.foe.co.uk/sites/default/files/downloads/overconsumption.pdf.
[8]Ellen MacArthur Foundation, 2014. www.ellenmacarthurfoundation.org.
[9]Webster, K., Bleriot, J., Johnson, C., 2013. A new dynamic: effective business in a circular economy. Ellen MacArthur Foundation, UK, ISBN:978-0-9927784-1-5.

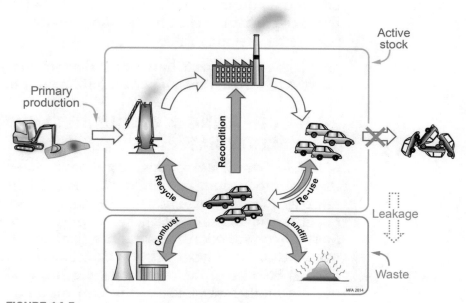

FIGURE 14.5
The circular materials economy. The aim is to retain materials in the "Active stock" box by reuse, reconditioning and recycling, minimizing leakage into the Waste box.

can be incinerated with energy recovery; only the remaining residue goes to landfill. The more material that can be retained within the green "Active stock" box the upper part of Figure 14.5, the less needs to be added through primary production. Indeed if design improvements use materials more efficiently it may be possible to function, at least for a while, with no primary production at all thereby creating a circular materials economy.

The *circular economy* means more than just efficient recycling. Taken literally, it means relying on renewable energy, tracking materials through the economy so that their location is known and using them in designs that allow their reuse with as little reprocessing as possible. The concept goes beyond the mechanics of production and consumption of goods, moving from the idea of *consuming* of materials to one of *using* them, somewhat in the way properly managed land is used for agriculture without consuming it. This implies a different approach to design, one that retains or regenerates materials during several manufacturing cycles[10] and reduces

---

[10]http://www.ellenmacarthurfoundation.org/.

demand on the use of critical materials. It is an important part of a resource-efficient, low-carbon economy, reducing costs and supply risks, and generating value.[11]

That is the ideal. Ideals have value; they set a target, something to be worked towards even if perfect fulfilment is not possible.

14.5   CREATING A CIRCULAR MATERIALS ECONOMY[12]

Material efficiency means providing more material services with less material production. Remember that material *production* is not the same as material *consumption*. Material production measures the quantity of materials created per year from ores, feedstock and energy. Material consumption measures the quantity of materials used in manufacture of all types per year. Consumption is greater than production because of the contributions of material recycling and reuse. Put another way, material efficiency means maintaining enough material in the Active Stock box of Figure 14.5 to provide the services we need while minimizing new material production flowing in from the left. Here we explore four broad strategies for achieving it:

- Better stuff: improved material technology
- Better design and longer product life
- Better business models
- Better behavior: regulation and change of life-style

14.5.1   Better Stuff: Improved Materials Technology

Improved material extraction and yield. Extracting and refining materials consumes 20% of global energy per year. Even an improvement of 1% in the efficiency of these processes would save much energy. New process routes offer potential for improved efficiency and greater yield. We need them: the ores we will mine in the future will be leaner than those we mine today.

New and improved materials. Material development, a century ago, was largely empirical. New materials emerged by trial and error

---

[11]https://www.innovateuk.org/competition-display-page/-/asset_publisher/RqEt2AKmEBhi/content/resource-efficiency-new-designs-for-a-circular-economy.
[12]The brief case studies that illustrate this Chapter are drawn from a number of sources, among them: WRAP, 2014. http://www.wrap.org.uk/content/innovative-business-models-1#k. Cradle to cradle case studies, 2014. http://www.c2ccertified.org/innovation_hub/innovators. Ellen Macarthur Foundation, 2014. http://www.ellenmacarthurfoundation.org/case_studies.

and lucky accident. The rapid expansion of materials science in the second half of the twentieth century provided a basis for reasoning about material development, but material behavior is sufficiently complicated that a component of inspired guessing and serendipitous discovery remained – an approach that has been described as "science-informed empiricism". Today, the science has evolved to the point that, coupled with contemporary information-processing power, *computational material design* has become a reality, opening a new era in material development.

The expansion of the world of materials. Strength and density, plotted here, are key properties of materials for transport systems. Figure 14.6 illustrates the dramatic expansion in the available range of these properties over the last 120 years. Materials for other applications have followed similar evolutionary paths. But with this gain in performance has come an increase in complexity – Table 1.2 gave a number of examples – and with it increased difficulty of recycling.

14.5.2   Better Design

Design for longer product life. A product reaches the end of its life when it is no longer valued. The cause of this rejection is, frequently, not the obvious one that the product just stopped working. The life expectancy is the least of[13]

- The *physical life*, meaning the time in which the product breaks down beyond economic repair;
- The *functional life*, meaning the time when the need for it ceases to exist;
- The *technical life*, meaning the time at which advances in technology have made the product unacceptably obsolete;
- The *economic life*, meaning the time at which advances in design and technology offer the same functionality at significantly lower operating cost;
- The *legal life*, the time at which new standards, directives, legislation or restrictions make the use of the product illegal;
- And finally the *desirability life*, the time at which changes in taste, fashion, or aesthetic preference render the product unattractive.

---

[13]This list is a slightly extended version of one presented by Woodward, D.G., 1997., Life-cycle costing. Int. J. Project Manage. 15, pp. 335–344.

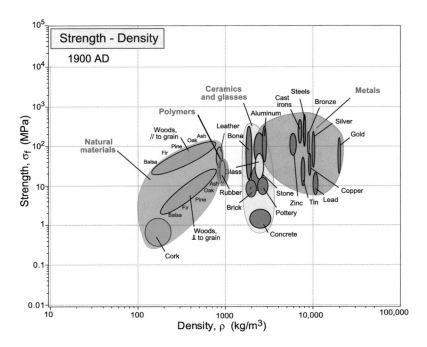

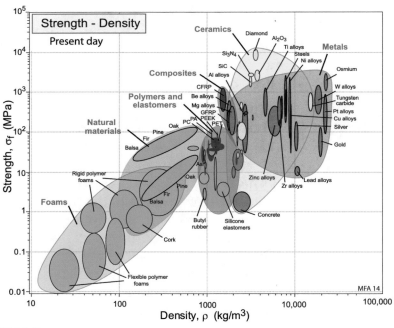

FIGURE 14.6

The dramatic increase in the strength and density of engineering materials between 1900 (on the left) and the present day (on the right).

One obvious way to reduce resource consumption is to extend product life by making them more durable – doubling the life, you might think, halves the consumption. But durability has more than one meaning: we have just listed six. Design to extend product life starts with an analysis of the life-limiting agency; there is little point in extending physical life (as an example) when the cause of rejection is outmoded functionality or change of fashion.

Most domestic appliances (refrigerators, washing machines, microwave and electric ovens) use much more energy during their life than it takes to make them in the first place. When this is so, extending product life may save materials but it does not always save energy. Advances in technology can reduce the use-energy; then making the product last longer can increase energy consumption. If the technology is advancing rapidly, early replacement can be more energy-efficient than extension of life. And where is the incentive for a maker of products to extend life when a longer life means selling fewer products? The answer to that may lie in the product-service business model. We come to that in a moment.

Design for reuse, repair and recycling. Products, when new, are worth much more than the materials they contain.[14] Value is added during manufacture and assembly; a family car, for example, is worth some six times more than the steel, rubber and glass of which it is made when it reaches the showroom (Figure 14.7, left hand side). After 8 years use, the car has depreciated by a factor of five or more but – if it is still usable – some residual value remains. More is lost if the car is broken down into components, and more still if it is shredded to allow its materials to be recycled (Figure 14.7, right hand side). Thus the sequence "reuse – repair – recycle" describes a descending path on the value chain: the first retains the most value, the last the least. But all three have the merit that the materials remain part of active stock.

Reuse. Products or their components can be taken back to be reused or restored for resale, a kind of non-destructive recycling. The user may change (1-year-old mobile phones are resold to those

[14]Green Alliance, 2014. Inside track. J. Green Alliance, UK. www.green-alliance.org.uk.

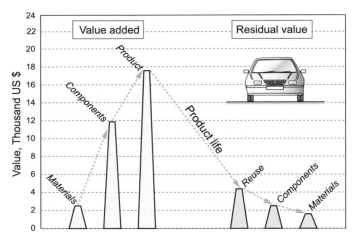

FIGURE 14.7
Products are worth much more than the materials inside them. Value is lost by breaking products down into components or materials.

who cannot afford a newest model) and the function may change (old school buses are re-used as camper vans, shipping containers are reused for housing) but their materials remain in active service. Well-established systems are in place to resell cars, boats, and houses for reuse when the first owner no longer wants them. For reuse to be a viable option there has to be a secure supply and consumers must trust it, needing quality assurance and price transparency. The success of eBay as an agent to enable reuse is a model for how this can be achieved.

Mazuma Mobile. The average first life of a smart phone is 18 months, after which time they end up in a drawer, hibernating. When the drawer fills up the phone goes to landfill. Mazuma targets these handsets early on to maximize their potential for reuse by reaching out to the customers directly, offering money in exchange.

"We try [...] to get the handsets as soon as someone upgrades so that we can maximize the amount of use of the handset while it is still current and still a good phone" says the CEO, Charlo Carabott.

The challenge was to gain public awareness, which Mazuma tackled by investing heavily in marketing and building trust from phone owners. Most of the 100,000 phones received each month can be reused after minor refurbishing, handled by an external partner. Phones that do not work are recycled to recover materials.

Repair and re-manufacture. Re-manufacturing is the upgrading or renewal of an existing product. Extending product life by repair makes business sense where products are expensive and labor is cheap but the business case for repair is weaker in developed economies where the reverse is true. There are many successful examples: the reconditioning of car and truck engines, the re-use of modules in photocopiers, the re-use of printer cartridges, and the re-treading of tyres. Re-manufacture works for products that are at the mature end of their life cycle, in a market in which technology change is slow.

Re-manufacture needs appropriate design to allow:

- Easy access and separation to allow fast, cheap disassembly;
- Easy identification of components via the model and assembly numbers; and
- Easy verification of condition by using embedded monitors to record use-history and predict residual life.

That all sounds good, but it is not so simple. Storage, administration and part replacement all have costs, and it is more difficult to automate repair than first manufacture.

Renault. The circular economy framework emphasizes design for remanufacture rather than aiming solely to use less material. The Renault automotive company has embraced the concept of a circular materials economy. The company's plant in Choisy-le-Roi, near Paris, remanufactures automotive engines, transmissions, injection pumps and other components for resale. More broadly, the company redesigns certain components to make them easier to disassemble for closed-loop reuse, enabling components from worn-out vehicles to become inputs for new ones. Renault reports economic benefits, including tighter control of its raw materials throughout its vehicles' life cycles. The plant remanufactured around 200,000 vehicle components in 2012 with a turnover of 100 million Euros.

Caterpillar Inc. Some products fit a circular business model more easily than others. Machinery, engines, gearboxes and drive trains, particularly, lend themselves to reconditioning and remanufacture. Caterpillar is a US manufacturer of construction and mining equipment. The company runs 'Cat Reman', a remanufacturing program that returns products at the end of their lives to as-new condition and seeks new ways to reduce, reuse, recycle, and reclaim materials that once would have gone into a landfill. Cat Reman took back over 2.2 million end-of-life units for remanufacturing in 2012. The company estimates that 65% of their costs are material-related so salvaging components makes business sense.

"Using less material may allow you to sell that first unit more easily and cheaply – but it does not form a relationship with the customer in the long term for reduced owning and operating costs" says Matt Bulley, Managing Director.

Historically, maintenance programs have been used to intercept products before they break. Caterpillar has gone further by installing digital links to units in the field allowing them to the company to monitor product status and intervene before damage or breakdown occurs.

Recycle. Recycling has one advantage over reuse. Reused components can only be used in the system from which they came; recycled materials, by contrast, re-enter the primary production cycle and thus can be used in a greater number of ways, even though the loss of value is greater.

The key steps are separation and recognition. Airframe makers now stamp sheet metal with the alloy designation and polymer molders imprint moldings with designated recycle marks. Recognition could be improved further by tagging components with bar codes or other identifiers that carry more detailed information, but none of these help if products are shredded before they are dismantled. To get further needs sophisticated, if unglamorous, materials science. When shredding is practiced, separation is possible by using the optical, magnetic, electrical, dielectric or inertial signatures of the materials, but this is imperfect, able to distinguish material families but unable to separate individual grades. What is needed is an internal identifier, pre-programmed by the manufacturer, which can survive life and carry a designation-tag to every shredded chip.

Ricoh. Ricoh, a global maker of office machines, designs its GreenLine brand of office copiers and printers to maximize component reusability. Its R&D department utilizes a lifecycle design approach to ensure that products can be easily dismantled for reconditioning and material recycling. To do this, Ricoh has set up recovery services that are comprehensive and easy to use, allowing recovery levels well above legal requirements. Mixing new and reconditioned equipment enhances Ricoh's portfolio with lower-cost models, serves a wider range of customers and generates additional revenue and margin by selling equipment more than once. It also retains materials as active stock in the use-cycle, contributing to resource conservation. Ricoh's objectives are to reduce the input of new resources by 25% by 2020 and by 87.5% by 2050.

Obstacles to full recovery. Reuse, repair or recycle of a given product line means collecting the products at end of first life. A truly circular system means collecting them all, but that is seldom possible. When materials are made into products, distributed, sold and used, there is what you might call entropy of dispersion. Think of recovering mobile phones at the end of first life. Most still work, and those that do not contain components and materials with value. All have value, but to recover it the phones have to be collected. Some are easy to collect but recovering every single one is unrealistic: the cost of the search would outweigh any gain. This trade-off can be modelled (see box below), with the result shown in Figure 14.8. There is an up-front capital cost to set up the recovery system. It is defrayed as units are collected and their value recovered. But the more that are recovered the harder it is to find those that remain, driving up collection costs. Figure 14.8 makes the point that 100% circularity is impractical; indeed the economically most attractive fraction – here

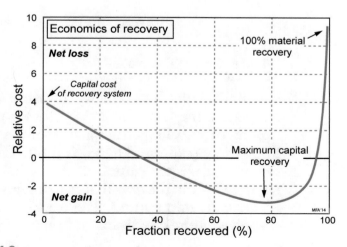

FIGURE 14.8
The relative cost of recovery as a fraction of fraction recovered. The economic sweet spot is at the minimum.

around 78% – can be much less. High direct (labor) cost shifts the minimum in the curve to the left; high residual value of the product (or of its materials) shifts it to the right. Appropriate legislation (penalties for discarding, or financial incentives such as buy-back) artificially shifts the minimum towards high recovered fractions.

Modelling recovery cost. Consider the broad economics of recovering units of a product to allow them to be reused, repaired or recycled. The capital cost of recovery system is C_c. The total number of units in circulation is n_{tot}. The number recovered at any point in time is n, so the recovered fraction is n/n_{tot}. The average value of a recovered unit is c_o. The value of the recovered fraction is then $n_{tot} c_o (n/n_{tot})$. The time t that it takes to recover a unit when n have already been recovered increases because of the dispersion:

$$\frac{dn}{dt} = \frac{1}{t_o}(n_o - n)$$

where t_o is a constant (the time to collect the first unit). Integration gives the recovery time per unit, t_r, as a function of fraction recovered, n/n_{tot}:

$$t_r = n_{tot}\, t_o\, ln\left\{\frac{1 - 1/n_{tot}}{1 - n/n_{tot}}\right\}$$

The direct cost of recovery is the recovery time t_r multiplied by the cost of time (labor, overheads etc), C_L. Summing the contributions gives the recovery cost C as a function of fraction recovered n/n_{tot}:

$$C = C_c - n_{tot}\, c_o\left(\frac{n}{n_{tot}}\right) + C_L\, n_{tot}\, t_o\, ln\left\{\frac{1 - 1/n_{tot}}{1 - n/n_{tot}}\right\}$$

This is the function plotted in Figure 14.8.

Circularity metrics. Can we measure circularity? Materials are retained in the "active stock" box of Figure 14.5 by reuse, reconditioning or recycling. The fraction of material that is not returned in this way (expressed as %) is plotted for a range of products as the vertical axis of Figure 14.9. About half the PET water bottles are recovered – the fraction that is lost per service cycle is about 50% (it depends on country). The lead in car battery, by contrast, so conserved more effectively; only about 5% is lost per service cycle.

The horizontal axis shows the service life in years. Newspapers have a useful service life of little more than 24 h. For PET bottles it is

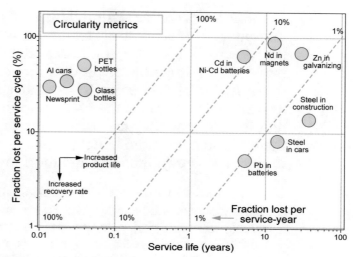

FIGURE 14.9
Circularity metrics.

closer to a week. The average life of a lead–acid car battery is about 5 years. Thus a more interesting metric is the loss per year of active service provision – the ratio of the two quantities plotted here. It appears as the diagonal contours on Figure 14.9. It varies dramatically, from less than 1% for structural steel to more than 100% (a factor of 2) for very short-lived products. The plot brings out the way in which increasing recovery rate and increasing product life slow the drain of active material stock.

14.5.3   Better Business Models

Replacing products by services. The replacement of goods by services increases use-intensity. Taxis are more intensively used than private cars. Laundromats are more intensively used than home washers. This product–service model can also bring increased customer loyalty and reduced marketing costs. From a materials perspective, service design means that the ownership of products and their materials remains in the hands of the product-maker. It is then in the interest of the maker to make the product last as long as possible and to allow reuse of its parts rather than scrapping them. This, in turn, drives higher specifications of design and choice of materials that increase life and durability.

Here are some much-cited examples. Xerox offers complete reproduction services instead of just selling photocopiers. Rolls-Royce sells "power by the hour" rather than aircraft engines. Schindler elevators sell "carefree vertical transport" instead of elevators. Dow Europe sells the services of chemicals instead of selling the chemicals themselves. GE Capital leases aircraft. These companies sell performance; it is not the product that carries the guarantee, it is the provision of the service. It is a small step from there to the idea of leasing the materials themselves, so that a manufacturer pays a rent for the time they are used but with the obligation to return them at the end of the rental period. Some materials are already treated in this way: uranium and thorium are examples.

Digital Lumens. Industrial lighting is inefficient and expensive. Warehouses, manufacturing, cold storage, and maintenance depots are in use 24/7, have high ceilings and require continuous illumination. Digital Lumens offers intelligent lighting systems that combine LEDs, networking, sensors and software to reduce lighting-related energy use by up to 90 percent. The LEDs have low wattage, long lifetime, little maintenance and are remotely controllable. The installation automatically adjusts lighting within a building by dimming or brightening bulbs to match the conditions. Digital Lumens now offers their lighting product as a service rather than an equipment sale. This service model has the appeal that it reduces the upfront investment and makes both the costs and benefits explicit.

Michelin. Tire-maker Michelin offers effective tire leasing programs with payment on a price-per-mile basis. The company started to lease automobile tires in the 1920s. Today, Michelin Fleet Solutions has 290,000 commercial vehicles in 20 countries under contract. The program offers reduced business risk, avoiding variability in tire performance, irregular purchase costs and unpredictable damage rates that can lead to swings in costs and negative cash flow.

All this sounds good, but it is not so simple. Here (paraphrased) are some revealing observations by the Sustainability Director of Interface,[15] a carpet company.

A failed experiment. "Product-service systems look like a panacea for sustainability. At Interface we put effort into leasing carpet, but it didn't take off. Here's why.

[15]http://greenallianceblog.org.uk/2011/09/29/can-services-replace-products-the-myths-and-truths/.

A carpet is not the whole functional unit. When you lease a car, it includes the service, road tax, etc. so the client receives the full service and doesn't need to do anything but enjoy the service. Leasing carpet is a bit like leasing a car wheel. You need the whole service, and in the case of carpet that is carpeted office space. There are already many people out there renting/leasing office space.

For a bank to finance a lease you need residual value. A £30k car after 3 years is still worth £10k–£15k. But carpet has negative residual value (cost of de-installation plus landfill or recycling). Which bank would be crazy enough to finance that?"

14.5.4   Better Behavior: Regulation, Social Adaptation and Change of Life-Style

Regulation. Governments have a broad range of economic instruments – carrots and sticks – to change the way we behave – we encountered some of them in Chapter 4, Section 5.6 and Figure 5.7. Such interventions may not be popular, but they can be effective. Most people would agree that reducing road deaths, improved public health and creating a stable financial system are desirable goals. But it took legislation to enforce the wearing of seat-belts and crash helmets, to ban smoking in public places and to make high-street banks back loans with adequate resources to make them happen. All three enactions triggered vocal opposition. But once in place they have achieved the desired outcomes and are now accepted. So regulation works. But as mentioned earlier regulation is also a wake-up call. It points to a design failure, something that better design could have made unnecessary in the first place.

Replacing materials by information. The applications on your computer probably cost more than the computer itself. The computer itself is made from materials but with it you can communicate, write, draw, do research, buy things and be entertained without using any material at all. At the commercial level information technology allows customized products to be made when ordered, removing the need to hold large stock. Similar dematerialization is spreading in retailing, manufacture, banking, education and entertainment.

Table 14.2 The Data-Intensity of Data-Objects	
Data Object	**Number of Bytes**
A page of text	2 kB
Range for PDF files	100–800 kB
A high resolution image	5 MB
An average Web site	5 MB
A video or audio download	1–10 MB
A full length movie	1–2 GB
Average i-phone user, per year	12–19 GB

The energy and carbon footprint of information. The "paperless office" may save materials but it still consumes resources. The digital economy (known as the ICT – the information-communications-technologies ecosystem) today consumes 1500 terawatt-hours per year, about 10% of the world's total electricity generation. Data-centers consume the most; between 10 and 20 kWh (36–72 MJ) of electrical energy per gigabyte of storage.[16] Table 14.2 lists typical file-sizes, from which the energy and carbon footprint of information-sharing can be calculated.

[16]Mills, M.P., 2013. The cloud begins with coal – an Overview of the electricity used by the global digital ecosystem. http://www.tech-pundit.com/wp-content/uploads/2013/07/Cloud_Begins_With_Coal.pdf?c761ac.

Doing with less, doing without. Rich economies consume more resources than are necessary to achieve an adequate standard of living. In developed countries cars, television sets, hi-fis and i-pods are seen not as luxuries but as necessities. Wealth in many communities is linked to status; possessions are wealth-made-visible. There is every sign that as the poorer but more populous countries become more affluent they will take the same view. Breaking the link between status and possessions brings a drop in material consumption. Some communities and religious orders have made this choice, electing to live without amassing more possessions than are necessary for life, but they are a tiny minority.

14.6   SUMMARY AND CONCLUSIONS

A linear industrial economy extracts natural resources, adds value by converting them into goods, distributes them, then rejects them as waste at end of life. For most of the nineteenth and twentieth centuries the prices of resources have steadily decreased in real terms, providing

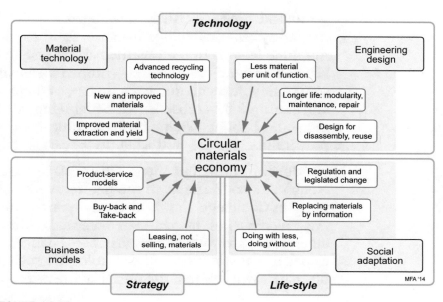

FIGURE 14.10
Contributions to creating a circular materials economy.

little incentive to conserve them. Since the beginning of the twenty-first century the trend has reversed; materials and energy have become steadily more expensive and increased demand from populous nations with rapidly growing economies has put stress on supply. A circular economy focusses not on acquiring resources but on conserving them by managing existing stock, maintaining it in active use and minimising the "leakage" of stock to waste. There are many ways to do so, some within the remit of the Materials Scientist and Engineer, some enabled by enlightened design to maximise useful life and ease reuse, others involving government incentives and controls, or requiring changed social perception and behaviour. The Chapter summarises these (Figure 14.10) and examines the obstacles and the opportunities they present.

A parallel is often drawn between natural ecology and industrial ecology, using the first as an exemplar of how the second might evolve into a balanced equilibrium with nature. But are natural systems really at equilibrium? Over long periods of time, yes. The forces for change are minimal, allowing optimization by natural selection at an ever more refined and detailed level. It leads to an interdependence on a scale that even now we do not fully grasp, but which man-made activities too frequently disturb. But on a geological time scale there have been massive disruptions of the natural system arising, for instance, from sudden climate change. Are there lessons in the way the natural system has then adapted? When dinosaurs – reptiles – succumbed to one of these disruptions, some small, furry mammal – a mouse, perhaps – survived, evolved and multiplied because it was better adapted to the changed climate and habitat. Where, in the technical world of today, are the post-industrial mice, and what do they look like?

We do not know. But there is a message here. The life raft, the Noah's Ark so to speak, of the natural world, allowing continuity in times of change, lies in its diversity: new mice, ready and waiting. The nuclei of the new system existed within the old and could emerge and grow when circumstances changed. How do we create a society – a scientific and technological society – with sufficient diversity to ensure that the nuclei of the next phase pre-exist in it?

14.7 EXERCISES

E14.1 How does a circular materials economy differ from the linear one largely practiced today? Make a two-box comparison like that of Table 14.1 highlighting the differences.

E14.2 Research the reuse of printer cartridges. What are the material savings? What are the economics? Most printer-makers have take-back schemes that provide useful information.

E14.3 What happens to used mobile phones? Research this question and write a short report on your findings.

E14.4 Recycling is practiced today – most metals are recycled to some degree (upto 50%), as is paper and glass. The recycling of plastics is still much more limited (2%). Why is this difference so large?

E14.5 Table 14.2 of the text listed the data-intensity of seven data-objects. Extend the table by two more columns and enter in them the energy-intensity (MJ_{oe}) and carbon footprint (kg $CO_{2,eq}$) of each. To do this, use the average electrical energy per GB of download cited in the text. Assume that the data-center from which the data are downloaded is located in Australia. You will find the carbon footprint of electrical energy in Table A10.6 in the Useful Numbers Appendix at the end of this book.

E14.6 Do you think manufacture without waste is possible? "Waste", here, includes low-grade heat, emissions and solid and liquid residues that cannot be put to a useful purpose. If waste is inevitable, why?

E14.7 What options are available for coping with the waste-stream generated by modern industrial society?

E14.8 Recycling has the attraction of returning materials into the use-stream. What are the obstacles to recycling?

E14.9 Car tyres create a major waste problem. Use the Internet to research ways in which the materials contained in car tyres can be used, either in the form of the tyre or in some decomposition of it.

E14.10 Packaging gets a bad press. It does, however, perform a number of useful functions. Research the essential and the useful, if not essential, functions of packaging.

Further Reading

Allwood JM, Cullen JM: *Sustainable Materials: With Both Eyes Open*, Cambridge, England, 2012, UIT, ISBN: 978-1-906860-05-9 (The authors present a carefully researched and documented case for increased material efficiency.).

Allwood JM, Ashby MF, Gutowski TG, Worrell E: Material efficiency, a White Paper, *Resour Conserv Recycl* 55:362–381, 2011 (An analysis of the need for material efficiency, possible ways of achieving it, and the obstacles to implementing them.).

Allwood JM, Ashby MF, Gutowski TG, Worrell E: Material efficiency: providing material services with less material production editors: *Phil Trans Royal Soc A* 2013, Number 1986, Royal Society, ISBN: 978-0-85403-999-9 (The proceedings of a discussion meeting devoted to this topic.).

Material efficiency: providing material services with less material production. In Allwood JM, Ashby MF, Gutowski TG, Worrell E, editors: *Phil Trans Royal Soc A*, vol. 371. London, 2013, Number 1986, Royal Society, ISBN: 978-0-85403-999-9 (The proceedings of a discussion meeting devoted to this topic.).

Allwood JM, Cullen JM, Cooper DR, Milford RL, Patel ACH, Moynihan M, et al: *Conserving Our Metal Energy*, Cambridge UK, 2011, The University of Cambridge, ISBN: 978-0-903428-30-9 (Ways of avoiding melting steel and aluminum scrap to save energy and carbon).

Azapagic A, Perdan S, Clift R, editors: *Sustainable Development in Practice*, Chichester, UK, 2004, John Wiley, ISBN: 0-470-85609-2.

Braungart M, McDonough W: *Cradle to Cradle: Remaking the Way We Make Things*, New York, NY, 2002, North Point Press.

Cradle to Cradle Products Innovation Institute. http://www.c2ccertified.org/innovation_hub/innovators, 2014.

Daly H: Toward some operational principles of sustainable development, *Ecol Econ* 2:1–6, 1990 (Herman Daly, an economist at the University of Maryland, offers a simple set of operational principles to guide sustainable development.).

Eggert RG, Carpenter AS, Freiman SW, Graedel TE, Meyer DA, McNulty TP: *Minerals, Critical Minerals and the US Economy*, Washington DC, 2008, National Academy Press (An examination of the criticality of 11 key elements, ranking them on scales of availability (supply risk) and importance of use (impact of supply restrictions)).

Ellen MacArthur Foundation: (The Ellen MacArthur Foundation has become an enabler of thinking and communication on the circular economy.) www.ellenmacarthurfoundation.org, 2014.

Ellen MacArthur Foundation: *Towards the Circular Economy: Economic and Business Rationale for an Accelerated Transition*. www.ellenmacarthurfoundation.org, January 2012.

Ellen MacArthur Foundation: *Towards the Circular Economy: Opportunities for the Consumer Goods Sector*. www.ellenmacarthurfoundation.org, January 2013.

Geiser K: *Materials Matter: Towards a Sustainable Materials Policy*, Cambridge, Mass, 2001, The MIT Press, ISBN: 0-262-57148-X (A monograph examining the historical and present-day actions and attitudes relating to material conservation, with informative discussion of renewable materials, material efficiency and dematerialization.).

Granta Design: "*Enabling Product Design and Development in the Context of Environmental Regulations and Objectives*" a White Paper. http://www.grantadesign.com/emit/, 2013.

Green Alliance: Inside track, *J. Green Alliance, UK*, 2014. www.green-alliance.org.uk.

Gutowski TG, Sahni S, Boustani A, Graves SC: Remanufacturing and energy savings, *Environ Sci Technol Am Chem Soc*, 2011.

Hertwich EG: Consumption and the rebound effect: an industrial ecology perspective, *J Ind Ecol* 9:85–98, 2005 (The rebound effect is the increase in consumption that occurs when improved technology reduces the energy consumption of products, cancelling the expected environmental gain.).

Lovelock J: *Gaia, a new look at life on earth*, Oxford, UK., 2000, Oxford University Press, ISBN: 0-19-286218-9 (A visionary statement of man's place in the environment.).

MacKay DJC: *Sustainable Energy – Without the Hot Air*, Cambridge, UK, 2008, UIT Press, ISBN: 978-0-9544529-3-3 (MacKay brings a welcome dose of common sense into the discussion of energy sources and use. Fresh air replacing hot air.) www.withouthotair.com/.

McDonough W, Braungart M: *Cradle to Cradle, Remaking the Way We Make Things*, New York, 2002, North Point Press. ISBN:13 978-0-86547-587-8 (The book that popularized thinking about a circular materials economy.).

Nguyen, Stuchtey M, Zils M: Remaking the industrial economy, *McKinsey Quarterly*, 2014. February issue. (A readable introduction to the ideas behind a circular materials economy) available on www.ellenmacarthurfoundation.org.

Schlösser T, Gayer U, Karrer G: *Technischer Bericht 0003–98*, Daimler-Chrysler Gmbh, 1999 (The Daimler-Chrysler Corporation have explored a number of natural fiber reinforced composites for their vehicles; this report presents the data shown in the figure in this chapter.).

Smil V: *Making the Modern World: Materials and Dematerialization*, Wiley, 2013. ISBN:10: 1119942535, ISBN:13: 9781119942535.

Stahel WR: *The Performance Economy*, 2nd ed., Basingstoke, Hampshire, 2010, Palgrave Macmillan.

USGS: *Mineral Commodity Summaries 2010*, U.S. Geological Survey, 2010 (The ultimate source of data for mineral production and metal production.).

von Weizsäcker E, Lovins AB, Lovins LH: *Factor Four: Doubling Wealth, Halving Resource Use*, London, UK., 1997, Earthscan publications, ISBN: 1-85383-406-8. ISBN:13: 978-1-85383406-6. (Both von Weizsäcker and Schmidt-Bleek referenced above, argue that sustainable development will require a drastic reduction in material consumption.).

Webster K, Bleriot J, Johnson C: *A New Dynamic: Effective Business in a Circular Economy*. UK, 2013, Ellen MacArthur Foundation, ISBN: 978-0-9927784-1-5 (A compilation of essays by distinguished authors discussing aspects of the Circular Economy).

World Bank: http://blogs.worldbank.org/prospects/category/tags/historical-commodity-prices, 2013.

World Economic Forum: *Towards the Circular Economy: Accelerating the Scale-Up across Global Supply Chains*, January 2014. available on www.ellenmacarthurfoundation.org.

WRAP: http://www.wrap.org.uk/content/innovative-business-models-1#k, 2014.

CHAPTER 15
Data, Charts and Databases

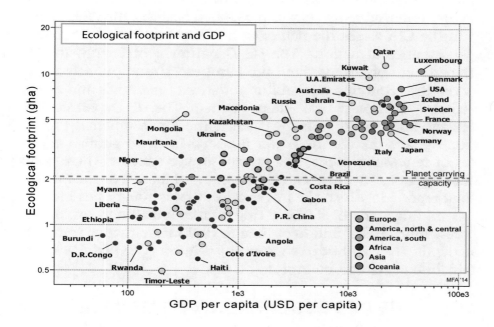

Chapter Outline

15.1   **Introduction and Synopsis**   242
15.2   **The CES Sustainability Database**   242
15.3   **Using the Elements Data-Table**   244
15.4   **Using the Materials Data-Table**   246
15.5   **Using the Power Systems Data-Table**   247
15.6   **Using the Energy Storage Systems Data-Table**   249
15.7   **Using the Legislation and Regulations Data-Table**   250
15.8   **Using the Nations of the World Data-Table**   252
15.9   **Summary and Conclusions**   257

Materials and Sustainable Development. http://dx.doi.org/10.1016/B978-0-08-100176-9.00015-3
Copyright © 2016 Elsevier Ltd. All rights reserved.

15.1   INTRODUCTION AND SYNOPSIS

All the information used in the case studies of Chapters 8–13 is accessible from open sources, but it is dispersed. The United States Geological Survey and the British Geological Survey compile data for material production and countries of origin. National and international agencies such as the United Nations, the World Bank and the CIA assess the demographic, economic and governance performance of nations. The US Department of Energy, the US Department of Defense, and the European Union publish analyses of the economic role and importance of materials and assign "critical" or "non-critical" status to them. The US Department of Energy, International Energy Agency, the Environmental Protection Agency, the Electric Power Research Institute publish analyses of renewable energy and energy storage systems. There is no shortage of trustworthy sources but many of them take the form of long reports in which the information that is sought lies hidden. Assembling some of it into a single, cross-linked network makes access much easier and frees time to explore alternative scenarios in depth. The Cambridge Engineering Selector (CES) EduPack Sustainability database attempts to do this.

15.2   THE CES SUSTAINABILITY DATABASE

The CES EduPack Sustainability database is a tool to help with Fact-finding. Figure 15.1 shows the structure and gives an idea of the content. It consists of a set of six linked data-tables.

The *materials* data-table (at the center of the figure) documents the compositions, properties and uses of over 3000 materials, including their environmental properties. It is linked to the *elements* data-table listing the properties of the elements of the periodic table and their countries of origin, annual production by country, environmental characteristics, material-criticality status and substitutability. The *legislation* data-table contains summaries of legislation, regulations, taxes and incentives to encourage or restrict the use of materials or of practices such as recycling that relate to material use. The *nations of the world* data-table contains records for the world's 210 countries, with data for population, governance, economic development, energy use and engagement with human rights, together with information that may bear on security

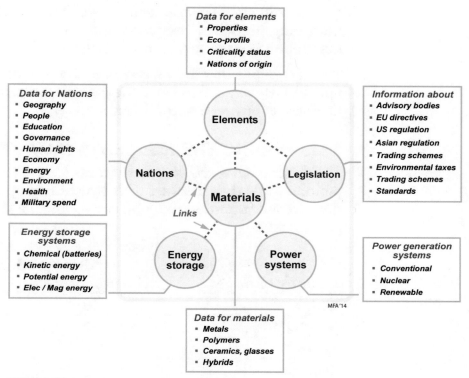

FIGURE 15.1
The structure and content of the CES sustainability database.

of supply and the ethical sourcing of materials. Two further data-tables provide information about *power generating systems*, and about *energy storage systems*. The database includes an eco-audit tool that allows a fast, approximate analysis of the energy requirements and the carbon emissions over the life-cycle of a product.

Links connect records in different data-tables that are in some way related. Thus the record for Stellite 6B (a Co–Cr–W alloy) in the materials data-table is linked to the elements cobalt, chromium and tungsten in the elements data-table, giving access to their countries of origin, annual production and criticality status. These elements records are linked in turn to records in the nations data-table for the countries from which they are sourced. Environmental restrictions in the legislation database are linked to the nations that have enacted them and the materials affected by them. The system contains further linked data-tables (not shown) for manufacturing processes, producers and references. A more detailed description is given in a pair of White Papers[1].

The paragraphs that follow give more information about the content and use of the database. These are demonstrated by using the database to answer a series of questions (*Qs*).

15.3   USING THE ELEMENTS DATA-TABLE

The *elements data-table* contains records for the 110 elements of the periodic table documenting their structure, properties, world production and countries of origin, environmental characteristics and critical status.

Country of origin	Indium production tonnes/year
China	340
South Korea	100
Japan	70
Canada	65
Belgium	30
Other	35
World	640

Q1. ***Material sourcing.*** *Indium is listed as a critical material. How much indium is produced each year and where does it come from? Why is it listed as critical?*

Answer. The record for indium in the elements data-table lists production and countries of origin. It is reproduced in the adjacent table. Over half of world production is from a single country. The record provides information about supply-risk under five headings, documented in the Section 15.8 of this chapter. Three – abundance risk, geopolitical risk and price volatility risk – are flagged as causes for concern. For these reasons it is classified as "critical" in both US and European lists.

---

[1]Ashby, M.F. and Ferrer, D. (2014) "Materials and Sustainable Development: Assessing Sustainable Developments – A White Paper" 2nd edition, and Ashby, M.F. and Vakhitova, T. "Material Risk and Corporate Sustainability – a White Paper" both published by Granta Design, Cambridge.

Critical elements	Grams per tonne in mobile phones
Platinum	70
Gold	140
Silver	1,300
Cobalt	19,000
Copper	70,000

Q2. ***Mining discarded electronics[2].*** *Mobile phones contain critical elements. The concentrations of five of these are listed in the adjacent table. Use the elements data-table to retrieve the typically mined ore grade for these elements. Do their concentrations in phones equal or exceed those of the ores from which they are currently extracted?*

Critical elements	Concentrations in phones, wt %	Typically mined ore grade, wt %
Platinum	0.07	0.00025
Gold	0.014	0.0018
Silver	0.13	0.055
Cobalt	1.9	0.5
Copper	7	2.6

Answer. The adjacent table repeats the concentrations of critical elements in mobile phones (expressed as wt %) and compares it with their concentration in their ores. The concentrations of all five elements in phones are larger than the typical grade of ore from which they are at present extracted. This suggests that "mining" waste electronics might provide a viable source of critical elements. There are, however, practical problems of collection, separation and refinement to be overcome.

Q3. ***Price and scarcity.*** *Are materials with low abundance in the earth's crust more expensive than those with high?*
Answer. Figure 15.2, made with the elements data-table of the Sustainability database, shows price and abundance in the

---

[2]http://www.electronicstakeback.com/wp-content/uploads/Facts_and_Figures.

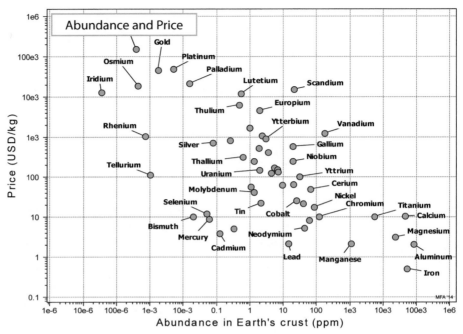

FIGURE 15.2
Abundance and price of elements.

earth's crust. There is scatter, but the trend is clear: materials with low abundance cost much more than those with high.

15.4   USING THE MATERIALS DATA-TABLE

The *materials data-table* contains records for over 3000 materials of engineering, documenting their properties, world production and environmental characteristics.

Q4. **Embodied energy.** *What is the "embodied energy" of a material? How large are these embodied energies? Which materials have low and which have high embodied energies?*

Answer Materials records list the embodied energies and carbon footprints associated with their production. Each is linked to text files explaining what the properties mean. The graphing facility in the software allows the data to be plotted. The adjacent figure shows a plot of embodied energies, allowing comparison (Figure 15.3).

Q5. **Eco-audits.** *Governments seek to minimize the energy consumption and carbon emissions of consumer durables such as washing*

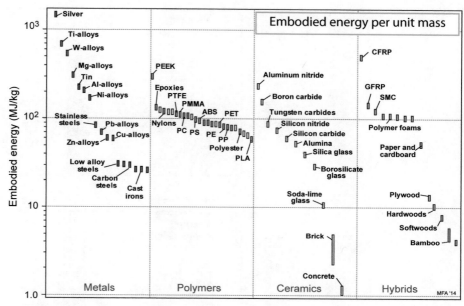

FIGURE 15.3
Embodied energies of engineering materials.

machines. *Over 21 million washing machines are sold in Europe each year[3]; globally, the number exceeds 60 million. The bill of materials and duty cycle of a typical washing machine are available. How can we identify the phase of life (material production, manufacture, use and disposal) that is most energy intensive and polluting?*

Answer. The eco-audit tool built into the database can help with this. The user enters the bill of materials, the distances transported and the duty cycle. Figure 15.4 shows the output of the tool. Although approximate, the result is unambiguous – the use-phase of the washing machine consumes far more energy and releases more carbon than all the others combined[4].

15.5   USING THE POWER SYSTEMS DATA-TABLE

Energy is central to any discussion of sustainability. The wish to de-carbonize electric power generation has stimulated interest in low-carbon and renewable-energy sources. Proponents claim that

---

[3]http://www.iseappliances.co.uk/index.php/aboutise/ise-news/43-26-million-washing-machines.
[4]For details of this and other audits, see, Ashby, M.F., (2013), "Materials and the Environment", 2nd ed., Butterworth Heinemann, Oxford.

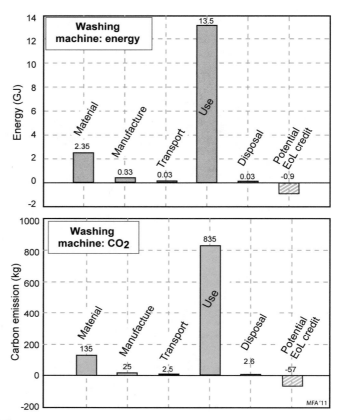

FIGURE 15.4
An eco-audit for a washing machine showing life energy and CO_2.

such systems are more "sustainable" than conventional fossil-fuel-based systems; opponents argue the opposite. The power systems data-table provides facts about power-generation systems in a consistent set of units.

Q6. *Carbon footprint of electrical power generation. Governments invest in alternative power systems to reduce dependence on imported fossil fuels and to reduce the national emissions of green-house gases, particularly carbon as CO_2. How do the carbon footprints of these alternative power systems compare with that of power from a conventional coal or gas-fired power station?* **Answer.** Figure 15.5 is made with the power generating systems data-table. It shows the sum of the carbon per kilowatt hour released by the fuel and that of construction and maintenance of the plant, pro-rated over a design

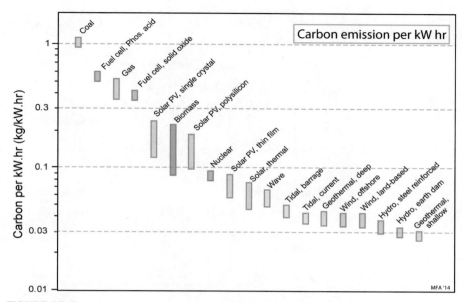

FIGURE 15.5
Carbon footprint of power from alternative power generating systems.

life of 20 years. All the power systems have a carbon footprint, but all are lower than that of conventional power systems. (This may not be quite fair. Coal and gas-fired power stations have a design life of 40–50 years. But for these the dominant source of carbon is the fuel, not the structure, so it makes little difference to the plot.)

15.6 USING THE ENERGY STORAGE SYSTEMS DATA-TABLE

Energy storage is important in two sectors.

- When energy is generated intermittently and its generation and consumption are not synchronized, the energy must be stored until it is needed. Grid-scale energy storage is a major challenge in deploying renewable energy systems, most of which are intermittent.

- Most transport systems carry the energy required to propel them. At present most such systems are driven by hydrocarbon fuels. Any attempt at decarbonizing transport faces the challenge of storing the necessary energy in a portable form.

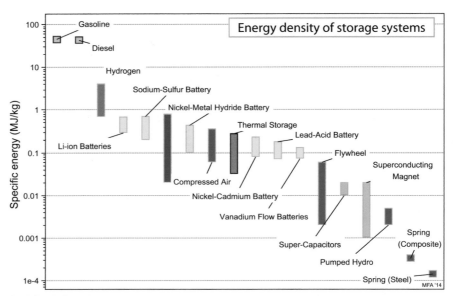

FIGURE 15.6
Energy density of energy storage systems. Diesel fuel and gasoline are show for comparison.

The energy storage data-table contains facts about energy storage systems in a consistent set of units.

Q7. *Energy storage systems for transport. Most transport systems carry the energy they need for propulsion with them. Most are powered at present by oil or gasoline. What alternative energy reservoirs might replace fossil fuels?*

Answer. Figure 15.6 is made with the energy storage systems data-table. It shows the specific energy of energy storage systems. The energy per kilogram of gasoline far exceeds that of any alternative system. Lithium-ion or sodium-sulfur batteries, candidates for electric car propulsion, are almost 100 times heavier for the same energy provision. Compressed air, attractive because it is clean and safe, has a still lower energy density.

15.7   USING THE LEGISLATION AND REGULATIONS DATA-TABLE

The *legislation and regulations data-table* summarizes legislation, standards, and taxation and incentive schemes that influence the use and disposal of products and materials.

End-of-Life Vehicles (ELV)

Summary of legislation. The European Community Directive, EC2000/53, establishes norms for recovering materials from dead cars. The initial target, a rate of reuse and recycling of 80% by weight of the vehicle and the safe disposal of hazardous materials, was established in 2006. By 2015 the target is a limit of 5% by weight to landfill and a recycling target of 85%. The motive is to encourage manufacturers to redesign their products to avoid using hazardous materials and to maximise ease of recovery and reuse.

Relevant sector. Automobile industry.

Enacting countries. European Union

Reference. ELV (2000) The EC 2000/53 Directive on End-of-life vehicles (ELV) Journal of the European Communities L269, 21/10/2000, pp. 34 - 42.

http://eur-lex.europa.eu/LexUriServ/LexUriServ.do?uri=CONSLEG:2000L0053:20050701:EN:PDF

FIGURE 15.7
A record in the legislation and regulations data-table. The records cover standards, legislation, regulation, taxation and incentives that impinge on the use of materials.

Q8. **European directives.** *What does the European End-of-life Vehicles (ELV) Directive say? Does it affect the choice of materials for a car?* **Answer.** The search function in the database allows a search on "ELV". Figure 15.7 shows the record: here, a summary of the European ELV Directive mandating recycling targets for vehicles. The materials database identifies materials that can and cannot be recycled. Meeting the 85% recycling target limits the use in vehicle production of materials that cannot be recycled. The materials data-table identifies these materials.

Q9. **European Directives.** *What does the WEEE Directive of the European Commission say?* **Answer.** A search on "WEEE" recovers a summary of the directive. It sets targets for collection, recycling and recovery of electrical products and is part of a legislative initiative to solve the problem of toxic contamination arising from waste electronic products. A link in the record gives direct access to the directive itself.

Q10. **REACH.** *What is the REACH Directive? Which nations apply this directive or one like it?* **Answer.** A search on REACH brings up seven relevant records. One is for the European Registration, Evaluation, Authorization and Restriction of Chemical Substances Directive (REACH). A second is for the Chinese equivalent known as China REACH. The other five are for legislation with a different name but similar intent. The directive places responsibility on manufacturers to manage risks from chemicals and to find substitutes for those that are most dangerous. Manufacturers in Europe and importers into Europe who use a restricted substance in quantities greater than 1 tonne per year must register the use. The list is long – it contains some 30,000 chemicals, many of them used in materials extraction and processing.

Q11. *Landfill tax. What is a land-fill tax? Approximately how much does it cost to send building waste to landfill in the UK?*
Answer. A *search* on "landfill tax" brings up the following information. A landfill tax is a tax on the disposal of waste. It aims to encourage waste producers to recover more value from waste, for example, through recycling or composting, and to use more environmentally friendly methods of waste disposal. The UK landfill tax in 2014 stands at £64 per tonne.

Q12. *CAFE rules. What are the US CAFE rules? Is there a European equivalent?*
Answer. A search on "CAFE" brings up the following information. The US Energy Policy Conservation Act of 1975 established the Corporate Average Fuel Economy (CAFE) standards, penalties and credits. The motive was to raise the fuel efficiency of new cars sold in the US from an average of around 15 mpg (miles per US gallon) to 27.5 mpg by 1985. The Energy Independence and Security Act, passed 22 years later (2007) raised the bar, aiming for a progressive increase to 35 mpg by 2020.

A further search for regulations relating to transport brings up both the CAFE rules and the equivalent European Regulation, the EU Automotive Fuel Economy Policy Directive (EC) No 443/2009 of the European Parliament, which sets emission standards for new cars.

15.8 USING THE NATIONS OF THE WORLD DATA-TABLE

The *nations of the world data-table*[5] contains information about the 210 independent countries of the world. Each record opens with a map and flag. This is followed by data for geography and population, indicators of wealth, well-being, economic development and respect for law and human rights. Each field-name is linked to an attribute-note explaining its relevance and source.

[5]The nations data-table is expanded from one first compiled by Professor Alex Wanner at the University of Karlsruhe.

Q13. *Wealth and Natural Capital. Globally, affluence is increasing. What influence does this have on the drain of Natural Capital?*
Answer. The nations data-table gives the following insights. The gross domestic product (GDP) per capita is one measure of affluence per person. The ecological footprint is a measure of the human impact on natural capital, per person. It is measured in "global hectares" (gha), thought of as the globally available productive land area per person if this productive land was shared equally. The productive capacity of the Planet can provide the present population (7 billion) with 2.1 gha per person. A nation with an average ecological footprint greater than 2.1 gha is consuming more than its fair share. Increased affluence, if shared and appropriately invested, has the potential to increase human and manufactured capital but the plot of ecological footprint against GDP per capita, shown as the cover picture of this chapter, strongly suggests that increased affluence causes a greater drain on natural capital.

Q14. *Wealth and Human Development. Does increased manufactured capital enable human development?*
Answer. The UN Human Development Index, a combined measure of health, education and welfare, is a measure of human capital. Figure 15.8 shows this index plotted against GDP per capita (a measure of manufactured capital). It would seem that the growth in per capita GDP up to about $5,000 (in 2008$) enables human development and with it human capital. Beyond that, the index flattens out.

Q15. *Wealth and Provision of Healthcare. Does increased manufactured and financial capital, measured by GDP per capita, increase human capital, as measured by life expectancy?*
Answer. Figure 15.9 is a plot of life expectancy against GDP per capita, made with the nations data-table. It shows that increased GDP increases at least one aspect of human capital – here measured by life-span.

Q16. *Press freedom and control of corruption. Does a free press help promote a society with low levels of corruption in Europe and North America?*
Answer. The nations records include the Reporters without Borders Press Freedom index and a number of measures of

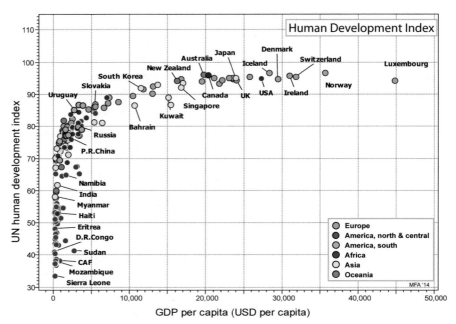

FIGURE 15.8
The UN human development index, plotted against GDP per capita.

the corruption, political stability and effectiveness of the rule of law. One of these is the World Bank Control of Corruption index. It is plotted against the Press Freedom index in Figure 15.10 for the countries of Europe, North and Central America. The correlation strongly supports the hypothesis that a free press helps support a just society.

Q17. *Corruption and human rights. At the time of writing Julian Assange, founder of Wikileaks, is resisting extradition to Sweden where he faces criminal charges and has sought asylum in the Ecuadorian Embassy in London to avoid it. In an article in the Times (Monday 20 August 2012), he is quoted as praising Ecuador for being a "courageous Latin American Nation that took a stand for justice." Research the justification of this statement by comparing the World Bank Control of Corruption index and the Reporter without Borders Press Freedom Index of Ecuador with those of Sweden. (Similar methods can be used to investigate nations from which materials are sourced.)*

Answer. The table shows the rankings for Sweden and Ecuador. Sweden ranks No. 3 for control of corruption and No. 12

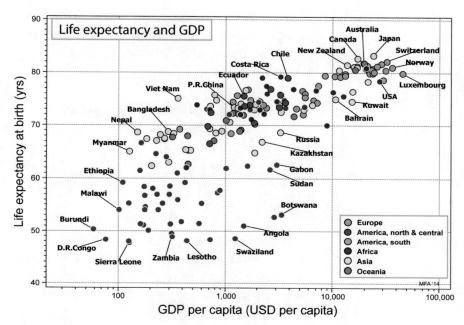

FIGURE 15.9
Life expectancy plotted against GDP per capita.

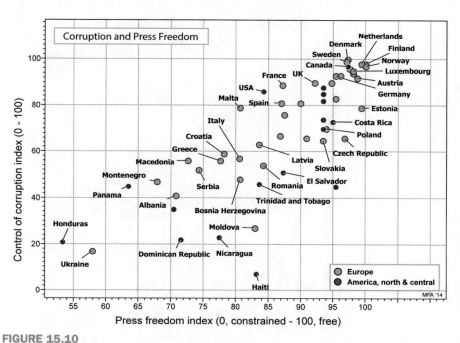

FIGURE 15.10
The World bank control of corruption index and the reporters without borders press freedom index.

for press freedom in the world. Ecuador, by contrast, ranks much lower. It appears that, by these two measures at least, Sweden has a rather better record of justice and free speech than Ecuador.

State	Control of Corruption Index (0 = poor, to 100 = excellent)	Press Freedom Index (0 = constrained, to 100 = free)
Sweden	99	97
Ecuador	20	68

Q18. **Global financial resources.** *Dealing with emerging global challenges – energy provision, provision of fresh water, emission reduction, desertification or climate change – will be expensive. To get the word "expensive" in perspective, use the data in the "THE WHOLE WORLD" record of the OTHER ENTITIES folder in the nations data-table to work out the amount that we, globally, spend each day defending ourselves from each other.*
Answer. Figure 15.11 shows the military spend per country per day. At the top left is the Whole World. Globally, we spend

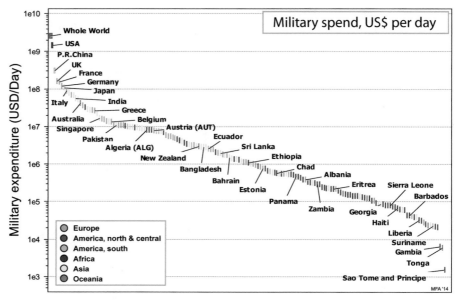

FIGURE 15.11
Military spend per day for the world and for individual nations.

about \$2.6 billion each day on defence. The positive side of this number is that, if the world were faced some major global threat, considerable resources could be diverted from the military spend to deal with it, if agreement could be reached to do so.

15.9 SUMMARY AND CONCLUSIONS

The CES EduPack Sustainability database is a resource to support the assessment of a proposed sustainable development. The database contains and connects information about materials and processes, energy, the environment, materials-related legislation and regulation, society and economics. The data it contains can be manipulated and plotted in ways that are helpful both in analysis and in presenting and explaining conclusions about a sustainable development. The numerous examples in this chapter illustrate some of the capabilities of the resource.

The database contains information about the criticality status of materials, about the governance and stability of the nations of the world and about legislation pertaining to materials, all useful background for exploring material-related risks.

Anticipating material criticality risk. The elements records of the database provide information about the nations from which materials are drawn and the supply-chain risk under five headings:

- *Abundance risk:* the risk to supply associated with the fact that a material is present in exceptionally low concentrations in the earth's crust and oceans.
- *Sourcing and geopolitical risk:* a measure of supply-chain concentration, the degree to which supply derives from one or a very few nations.
- *Environmental country risk:* the risk that environmental legislation enacted by supplying nations might disrupt supply.
- *Price volatility risk:* a measure of the fluctuations in the price of a material, calculated as the percentage difference between the maximum and minimum price (in USD/kg) over the past 5 years, relative to the minimum price.
- *Conflict material risk:* the risk that the sourcing or production of a mineral may have financed conflict and thus be blacklisted under the US Dodd-Frank Act and analogous European legislation.

The nature of the risk helps guide strategies to stabilize supply.

Anticipating and managing social risk. A company with off-shore operations must understand the state of the development, the weaknesses and the needs of the communities in which they operate. The nations of the world records provide this background information and help identify ways in which the company can contribute to the society with who they work. As examples, the nations records give information about literacy, gender-equality (or lack of it), median income, equality of wealth distribution, freedom of speech and level of corruption.

Anticipating economic risk. The costs of off-shore material production or manufacture is influenced by the cost of energy, water, labor and by the level of taxation. The nations records contain country-specific data for these.

Anticipating regulatory risk. All major nations have environmental taxes and regulate or restrict the products that can be sold and used within their borders. Products exported to a nation must comply with that nation's laws. The legislation and regulations records of the Sustainability database describe a number of these laws and act as a prompt in planning corporate strategy.

CHAPTER 16

Guidance for Instructors

Chapter Outline

16.1 Introduction and Synopsis 260
16.2 Problem-Based Learning 260
16.3 PBL and Sustainable Development 261
16.4 Organizing the Project; Scheduling the Activities 262
16.5 Assessment 266
16.6 Feedback from Students (UPC Course "Sustainable Design" 2012, 2013) 267
16.7 Summary and Conclusions 270
16.8 Suggestions for Further Projects 271

Materials and Sustainable Development. http://dx.doi.org/10.1016/B978-0-08-100176-9.00016-5
Copyright © 2016 Elsevier Ltd. All rights reserved.

16.1 INTRODUCTION AND SYNOPSIS

In recent years many universities have sought to incorporate concepts of sustainable development into engineering, materials and design programs. There are two ways of doing it: by creating new courses or by embedding it across the existing programme. The approach described in this book can help with both.

This chapter describes how the approach interfaces with problem-based learning (PBL) methods and the match between its learning objectives and the accreditation criteria of ABET, CDIO and EUR-ACE. It discusses student difficulties and how to overcome them, and reports the views of students who have graduated from a Masters level course that used the method.

16.2 PROBLEM-BASED LEARNING[1]

PBL is a learning strategy based on the acquisition of knowledge through realistic problem-solving. A small group of students supported by a tutor analyzes and solves a problem designed to attain certain learning outcomes. In the process students diagnose their learning-needs, work as a team to acquire the information they need, develop skills of analysis and synthesis of information and of communication both within and external to the group. The focus is on the student and the learning outcomes rather than on the instructor's knowledge and a formal curriculum. It differs from the "tell-and-test" approach by requiring students to identify what they need to know and then finding out how to acquire it. PBL aims to develop the following skills.

- *Knowledge and Understanding* – to find and use information, and to be able to use this knowledge in new and unfamiliar situations.
- *Skills and abilities* – to put knowledge and understanding to practical purpose.
- *Values and attitudes* – to use knowledge and understanding responsibly.
- *Critical thinking* – to develop a reasoned position.

[1]Hmelo-Silver, C. E. (2004). "Problem-Based Learning: What and How Do Students Learn?". Educational Psychology Review 16 (3): 235.

- *Systems thinking* – to integrate knowledge and values into their wider context.

Experience shows that barriers have to be overcome when introducing PBL methods.

- *It is a difficult transition.* Teachers and students have to change their vision of learning, assuming responsibilities for activities that do not appear in conventional class-room teaching.
- *Curricular modification.* Projects generally draw on a number of disciplines. It may be necessary to modify the teaching of these (in other courses) to make them compatible with the PBL approach.
- *Teaching skills.* Teachers may lack the training needed to work with student groups in the role of a coach rather than a lecturer. Considerable inertia surrounds the long-established methods of teacher-centered class presentations transmitting information in a one-way manner. In PBL it is the student and the learning that are center-stage. Acquiring techniques for managing group interaction (cohesion, communication, competition, etc.) can require practice.
- *More time is required.* It is not possible to transfer information as quickly by PBL as with conventional methods. More teacher-time is needed to prepare the problems, to supervise students and give feedback. But, it is argued, the learning is more deeply embedded.

The motivation for PBL is nicely captured in this quotation from Benjamin Franklin (1706–1790):

> *"Tell me and I forget, teach me and I may remember, involve me and I learn"*

16.3 PBL AND SUSTAINABLE DEVELOPMENT

Problem-based teaching with elements of PBL is a powerful medium for introducing issues of sustainability. The requirements for the student to formulate the objective clearly, to understand stakeholders' positions, to develop and use fact-finding skills to assemble relevant information and form a judgment of the significance of the facts presents challenges that a passive lecture-style of teaching does not do. The approach described here fosters a number of skills:

- Team and transdisciplinary working
- Systems thinking
- Creative and critical thinking
- Self-reflection
- Cognitive (knowing), functional (doing), personal (behaving) and ethics (valuing) skills

Today most academic programs are subject to accreditation standards that set quality requirements; among them are ABET, CDIO and EUR-ACE. Table 16.1 lists possible learning outcomes of a course on sustainable development, linking them[2] to the acquision of

- *Skills*. What students should be able to do by the time the course is completed
- *Knowledge*. What students should know and understand by the time the course is completed
- *Attitudes* (or *Values*). What the students' opinions will be about the subject matter of the course by the time it is completed

The Table flags the items of those programmes that correspond to the intended learning outcomes. The real actual outcome will, of course, depend on the specific activity. Three basic principles are followed:

- Understanding the context and the actors involvement of a problem.
- Understanding how to reconcile the conflict between the ideal and the reality.
- Recognizing that diverse points of view can co-exist and still allow a socially-acceptable solution.

16.4   ORGANIZING THE PROJECT; SCHEDULING THE ACTIVITIES

The methods described in this book support a problem-based teaching and learning approach. They can be used for both individual or group projects but work best with groups because of the learning that comes from interaction. The description below and

---

[2]http://web.mit.edu/tll/teaching-materials/learning-objectives/index-learning-objectives.html.

Table 16.1 List of Learning Outcomes, Their Level of Depth in Courses that Use the Five Steps Methodology, and Correspondence with Accreditation Standards

Learning Outcomes	Levels of Knowledge, Skills, Attitudes[a]						Accreditation Criteria		
By the Time the Student Finishes the Course they Should be able to.....	Knowledge	Compre-hension	Application and Analysis	Synthesis, Creativity, Evaluation	Receiving, Responding, Valuing	Compare, Relate, Synthesize	ABET	EUR_ACE	CDIO
1. Know and understand the basic dimensions of sustainable development	•	•			•		a, j	1.2, 1.4	1.2
2. Understand the complexity that surrounds assessment of sustainable developments	•	•			•		f, h	1.5, 5.5, 5.7	2.3, 4.1
3. Conduct a systematic assessment of a proposed sustainable development			•		•		b, k	2.2, 2.4	2.2, 2.3
4. Synthesize multi-disciplinary information associated with a single problem			•	•	•	•	e	5.5	2.1
5. Identify stakeholders and their perspectives on a specific issue			•		•	•	j	5.4	4.1
6. Find information from diverse sources and form a judgement about its reliability			•		•		b	4.1, 4.4, 4.6	2.2
7. Present the results of their investigation clearly and susinctly					•		g	6.2	3.2
8. Work in teams on a specific project			•		•	•	d	6.1	3.1
9. Evaluate options for sustainable development technologies					•	•	h	2.4	2.3, 3.4

[a]Bloom, B. S.; Engelhart, M. D.; Furst, E. J.; Hill, W. H.; Krathwohl, D. R. (1956). Taxonomy of educational objectives: The classification of educational goals. David McKay Company, New York

the schedule in Table 16.2 outline the way we have tried to do it. The prompt-lists in Chapter 4 and the capitals/sectors matrix of Table 4.5 give help with each of the stages.

Step 1: Frame the objective and scope. Initially each group selects a problem, clarifies the objective and plans the development of the

Table 16.2   Scheduling the Activities

Phase	Activity	Pedagogical Observations	% of Time
Introduction	Review the goals of the course Introduction to the method Full example in class (simple). Use summary forms for each phase. Assign students to groups of 3–5.	Explain the role of the students as consultants for a committee seeking advice.	10
Case selection+Step 1. Goal definition	Project selection from an open list with goal background information on motivation. Student group clarifies objective and scope.	Encourage students to suggest topics to study. Consider having groups working in parallel on the same topic.	10
Step 2. Stakeholders analysis	Identify stakeholders and their main concerns with regard to the theme	Introduction to stakeholder analysis If time: arrange interaction with real stakeholders	40
Step 3. Fact-finding	Introduction to the CES EduPack and Sustainability database Use of CES for fact-finding Illustrate the use of the Internet for fact-finding	Locate data for Materials/ Design/Environment/Legislation/ Economics/Society	
Step 4. Synthesis/Relationships	Discuss and debate relationships of facts and the three capitals	Give an example of how to use the synthesis matrix	30
	Role playing: one group acts as "consultants" invited to present their preliminary results to a committee. The committee is composed of members of the other groups who are free to question facts and their interpretation. Revise the conclusions and incorporate new thinking.	Do not sacrifice this phase. Debate and role playing is fundamental for learning from all groups work, not just from their own project. Create a "credible" scene.	
Step 5 Reflection and alternatives	Assemble conclusions: Is the prime objective achievable? Are stakeholder concerns addressed? Submit full report and presentation.	Does the project qualify as a sustainable development? What related developments might be needed to make it so?	10

subsequent phases. It is essential that the students approach it as a systems problem, one with some aspects of rational analysis, others that are conditioned by culture, beliefs and by group interests and influences. Student motivation is greatest if they have chosen the problem themselves, but certain criteria should be met, namely that:

- Technology and materials play a role in its analysis.
- Its goal and scope are achievable in the time allotted to the course.
- Sources of information are available to quantify the background to the project and to explore stakeholders concerns.

Step 2: Stakeholders. The next step is to identify the relevant stakeholders and list their concerns, something unfamiliar to most students. Once identified, the roles of stakeholders can be assigned to individual group members. The individuals research their assignation, taking on the role of their assigned stakeholder, and report their concerns back to the group as a whole (Figure 16.1). Section 4.3 of the text, given as assigned reading, can help here.

Step 3: Fact-finding. Again, the task can be divided among the group-members, each exploring an aspect of the problem (*"Materials"*, *"Environment"*, *"Legislation"* etc.) and reporting back to the group as a whole. The Internet gives access to factual sources. Not all are reliable; key facts should be checked against a second, independent source. The CES EduPack Sustainability resource (Chapter 15) provides factual information about materials, energy, countries, legislation and, via the eco-audit tool it contains, the environmental impact of a product life-cycle. It cannot provide everything but it is

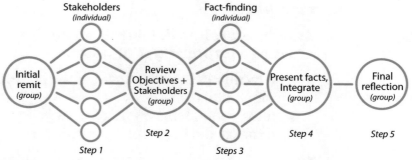

FIGURE 16.1
Running a sustainable-technology assessment project as a group activity.

a good starting point for traceable data. Section 4.4 of the text also gives support here.

Step 4: Synthesis – assessing the impact of the facts on the three capitals. The group is then in a position to debate the impact of the facts on the three capitals, seeking a consensus that recognizes the existence of more than one point of view. This is where judgement enters. The matrix that appears in each of the case studies of Chapters 8–13 can be copied, enlarged if necessary and handed out to students. Students find this step to be a difficult one. Debate and discussion to resolve divergent views is an important learning outcome. Section 4.5 of the text gives support.

Step 5: Reflection. This is the creative step, one that should include both near-term and long-term thinking. Are there new, possibly disruptive, solutions that achieve the prime objective but in a better way? Creativity aids can help here. Some are summarized in Section 4.9 of Chapter 4.

It is suggested that the final reflection is presented to the whole class. A peer evaluation among groups can be valuable: it obliges students to analyse critically the work of others and helps them to think about the analysis pattern of the five steps (the only commonality among the groups) beyond the details.

Throughout, students are encouraged to report the results in a simple manner, making them accessible to non-experts[3].

16.5 ASSESSMENT

There are many strategies and tools to assess learning when working with projects, among them, the *assessment rubric* and the *learning portfolio*.

Assessment rubric. A rubric is an explicit set of criteria used for assessing a particular type of work or performance.

- it facilitates the assessment of the project and it acts as a guide for grading;
- it makes students aware of what is expected of them; and
- it makes the five steps methodology easier to understand.

---

[3]Professor John Abelson (University of Illinois) raises an interesting question: do we need a higher level (systems and future development) in the synthesis step? Otherwise many options will seem closed or unlikely. Should "Synthesize" pose a set of issues to be resolved?.

Table 16.3 is an example of such a rubric. Three levels of quality (+, ≈ and –) are defined for each step. The percentage in the first column indicates the weight given to each phase in the final grade. It is up to the teacher to transform this into specific values.

Learning portfolios. A portfolio is a systematic collection of evidence that demonstrates what a student has learned over time. It should include both work samples and reflections on learning. The expectations of a portfolio are the following: clarity of goals, explicit criteria for evaluation, work samples tied to those goals, student participation in selection of entries, teacher and student involvement in the assessment process, and self-reflections that demonstrate the students' understanding of what worked for them in the learning process, what did not, and why. These elements enhance the learning experience and the self-understanding of the student as learner.

In the present context the portfolio should be assembled over the course of the project and should include the work done by the student for each phase of the analysis and their self-reflection about their own learning in that phase. This allows the instructor to react in real-time, providing feedback to steer the learning towards the expected outcomes (Table 16.1).

16.6   FEEDBACK FROM STUDENTS (UPC[4] COURSE "SUSTAINABLE DESIGN" 2012, 2013)

Opinion

- Consistent, holistic, very useful approach. The five step method is simple and concrete – a useful framework for tackling complex problems.
- The first step: "clarify objective" is essential but not easy. Defining the scale needs research because of the uncertain repercussions in the long term.
- The CES EduPack Sustainability database is easy to understand and use.
- The method concludes with a step to propose alternatives. This last step differentiates it from other analysis methods as it allows for changes in paradigms, a critical issue of

---

[4]Universitat Politecnica de Catalunya, Barcellona, Spain.

Table 16.3 Suggested Rubric for Marking Projects

Step	+	≈	−
Step 1 prime objective (5%)	▪ Prime objective is well defined ▪ Time scale is explicitly stated ▪ Physical scale is explicitly stated	▪ Prime objective is well defined	▪ Prime objective not well defined
Step 2 Stakeholder analysis (20%)	▪ All relevant stakeholders identified ▪ Concerns identified from sound sources ▪ Influence/power analysis competent ▪ Stakeholder summary well balanced	▪ Most relevant stakeholders identified ▪ Concerns are identified ▪ Stakeholder summary adequate	▪ Relevant stakeholders are missing, or ▪ Stakeholder concerns misinterpreted, or ▪ Stakeholder summary missing
Step 3 Fact-finding (25%)	▪ Facts well researched (for Materials, Design Environment, Regulation, Society, Economics) ▪ Sources are scientifically referenced ▪ Areas are organized in sub-areas	▪ Facts researched (for most of Materials, Design Environment, Regulation, Society, Economics) ▪ Sources are referenced	▪ Facts poorly researched. Significant areas omitted (Materials/manufacture, Environment, Design, Regulation, Society) or ▪ Sources are not referenced
Step 4 Synthesis (25%)	The impacts of the six areas are: ▪ Evaluated and justified for the three capitals ▪ Balanced and ranked for the three capitals	The impacts of the six areas are: ▪ Evaluated and justified for the three capitals	The impacts of the six areas are: ▪ Evaluated for the three capitals (if any) ▪ Not justified or balanced
Step 5 Reflection (25%)	Reflection, considering: ▪ If the Prime Objective can be achieved ▪ If stakeholder concerns are addressed ▪ Which capitals are augmented? Which are disadvantaged? ▪ Do benefits outweigh the negative impacts? ▪ Design alternatives are proposed in order to: ▪ Achieve Prime Objective in more equitable and innovative ways ▪ Consider possible achievement of Prime Objective on both short and long time scale	Reflection, considering: ▪ If the Prime Objective can be achieved ▪ If stakeholder concerns are addressed ▪ Which capitals are augmented? Which are disadvantaged? ▪ Do benefits outweigh the negative impacts?	Reflection omits ▪ To examine if Prime Objective can be achieved ▪ To examine if stakeholder concerns are addressed, or ▪ To consider the balance between the three capitals

sustainability. The data-table of national characteristics in the database is particularly helpful.

Learning – individual views

- I had not seen nature or society defined as a capital, and I was surprised about it. It broadened my vision about sustainability.
- I think that this method will be useful for me to explore different projects, even after having finished the Master's degree.
- I learned that it is better to have an order to follow when we want to obtain a good result in a project.
- The initial step [*clarify the objective*] encourages the user to think and research rather than take immediate action. The intermediate steps then involve gathering information before trying to form judgments.
- The most beneficial part of the process for me was the third step, fact-finding and its components.
- There is room for debate which I believe is a positive aspect.
- The CES database is a useful tool in assessing the actual sustainability of products considered to be sustainable. It and the method reinforce the fact that, regardless of the popular view of sustainable projects, all of its benefits and drawbacks must be weighted in order to measure just how sustainable it is.
- A phrase in the handout that caught my attention and summarizes this approach very well is: *"There is no completely "right" answer to questions of sustainability – instead there is a thoughtful, well-researched response that recognizes the many conflicting facets and seeks compromises that offer the greatest good with the least bad."*

Students suggested the following additional aids:

- Supply forms and summary tables to be filled in by the student.
- Provide more guidance on stakeholder analysis (now included in Chapter 4).
- As a group we struggled to define the prime objective and its scale – more help with this would be useful (now included in Chapter 4).
- Allow more time to get familiar with the CES software because it is particularly useful for the project. Perhaps an "index"

suggesting where to find each of the six categories of information (materials/energy/environment/legislation/society/economics). (Chapter 15 now includes this.)

■ Perhaps the process should be more iterative, allowing evaluation and reflection on each step to check that it is going as planned.

About the overall approach:

■ I really appreciated the examples given in class about how to apply the procedures. Do a full example of a project (such as the ones presented in the handout) in class to illustrate the method. Perhaps present it briefly at first, and then expand on it.

■ Different groups will likely develop different analyses of the same articulation. It would be interesting to compare the analyses of two groups that both tackle the same project.

■ There should be the *"Is this development really needed at all?"* question somewhere.

■ I would enjoy some group discussions, debates, or role plays where everybody gets a role that they have to read into and play in class, followed by some kind of debate. I think it would be both fun and educational.

16.7   SUMMARY AND CONCLUSIONS

Our experience of using the sustainable development analysis as the framework for student projects provided the following insights.

■ Students appreciate the methodology. It provides a structure that allows a systematic approach while remaining holistic and recognizing the inherent complexity of sustainability issues.

■ It helps to give one full example before starting the projects so that students see how the full process works (one of the case studies in the text could be used for this). It can also help to show failed examples like those described in Chapter 7.

■ Students tend to have difficulties formulating the primary objective and the size and time scale. Good support should be given when framing these.

- Students are generally unfamiliar with stakeholder analysis. Allocating time to explain the concept is important; grasping it is a valuable educational experience.
- Using the sustainability database of CES EduPack can save time and provide much relevant information. A short introduction to it (1 h) is useful.
- The matrix between the three capitals and the six sectors (Table 4.5 and its use in the case studies) helps students in the synthesis phase, one that they find difficult. Explain and discuss the issues valued in the three capitals and the relationships between them. Table 4.4 helps here.
- For the final presentation, it helps to create a close-to-real setting. It should take the form of a presentation to stakeholders. Role playing (or participation of the real stakeholders) is recommended.

16.8 SUGGESTIONS FOR FURTHER PROJECTS

Each of the case studies of Chapters 8–13 ends with suggestions for related case studies that draw on the background of the chapter itself. Here we list some further suggestions for case studies.

Element	Mass in a mobile phone (milligrams)
Platinum	0.077
Gold	0.15
Silver	1.4
Cobalt	21

P16.1 **Rare elements in mobile phones.** About 1.5 billion mobile phones will be produced in (2013). The table gives approximate values for the content of four rare metals in phones. Explore the supply-chain risk associated with one of these following the approach used for neodymium and lithium in the text.

   1. Assess supply-chain concentration, using data for countries of origin in the materials records of the sustainable development database.

2. Research the rule of law and the press freedom (a measure of freedom of speech) in the nations from which most of the metal is sourced.

3. Find the annual world production of the materials from the materials data-table and compare it with the annual demand created by phone manufacture, assuming none is recovered from old phones.

4. Explore the viability of recycling[5] rare metals from mobile phones.

P16.2 **Electric hand dryers versus paper towels.** Paper-based and electric hand dryers compete. Many people prefer the first because it is quick (about 8 seconds) and clean. But many providers prefer the second (drying time at least 45 seconds) because it requires less attention. Explore both, viewing each as an expression of a more sustainable technology than simply wiping hands on trousers.

The British designer James Dyson produces an electric dryer that dries hands almost as fast as paper towels (10 seconds). Is it a more viable articulation of sustainable technology?

Cloth towel	Paper towels	Conventional dryer	Dyson Airblade
Power 0 kW	Power 0 kW	Power 2.4 kW	Power 1.6 kW
Drying time ≈ 8 s	Drying time ≈ 8 s	Drying time ≈ 20 s	Drying time ≈ 6 s
Price of fixture ≈ £0.1	Price of fixture ≈ £20	Price ≈ £250	Price ≈ £700

P16.3 **Geothermal power for Britain[6].** Energy companies are in discussion about reducing Britain's dependence on gas and coal-fired electric power by piping electric power, generated

---

[5]Ellis T. W. Schmidt F.A and Jones L.L "Recycling of rare earths" http://www.osti.gov/scitech/biblio/10190438.

[6]The Times, November 12, 2012. http://www.theengineer.co.uk/home/blog/icelands-volcanoes-could-power-the-uk-but-at-what-cost/1012334.article.

in Iceland from geothermal heat, to Britain through an undersea cable. In theory it is possible to pump low-carbon electricity from Iceland to the UK to meet up to one-third of the UK's average energy consumption. The cable would be at least 1200-km long, each kilometer requiring 800 tonnes of copper. The cost is estimated at about £1,000,000,000 (one billion pounds). Does this technology make environmental and economic sense? Is it politically desirable? Carry out an assessment of this articulation of sustainable development.

P16.4 **Bio-composites as non-structural parts of vehicles.** A number of auto makers are trialing bio-composites for non-structural components such as inner door panels and interior trim. The prime objective is to replace materials derived from non-renewable resources (steel, oil-based plastics) with those that are renewable and (possibly) lighter. But does this articulation make sense? Are the properties of bio-composites as good as those they are supposed to replace? If they are not, sections will have to be thicker, so possibly heavier. Do bio-composites have a lower embodied energy than glass fiber reinforced sheet molding compound? Will consumers accept them (some smell a bit)? See what you can find out about them and then debate their contribution to the three capitals.

P16.5 **Alternative light-sources for traffic lights.** Every city and town of any size has traffic lights to control traffic flow. The energy they consume is significant. Light-emitting diode (LED) technology is now sufficiently advanced that they could replace tungsten filament technology globally. Explore this as a viable articulation of sustainable development, bearing in mind the demands this would place on the elements that are essential for LED technology. Is the present global supply sufficient to enable this development on a global scale?

APPENDIX

Useful Numbers

A.1 INTRODUCTION

If you want to be quantitative you need numbers. This chapter assembles numbers to help with analyzing sustainable developments. It starts with general physical constants and units. The six sections that follow present data relevant for the six sectors of the hexagon – Materials, Energy, Environment, Legislation, Society and Economics.

Table A.1 Physical Constants in SI Units

Physical Constant	Value in SI Units
Absolute zero temperature	$-273.2\,°C$
Acceleration due to gravity, g	$9.807\,m/s^2$
Avogadro's number, N_A	6.022×10^{23}
Base of natural logarithms, e	2.718

(Continued)

Table A.1 Physical Constants in SI Units—Cont'd

Physical Constant	Value in SI Units
Boltzmann's constant, k	1.381×10^{-23} J/K
Faraday's constant, k	9.648×10^{4} C/mol
Gas constant, $\bar{R}$	8.314 J/mol/K
Permeability of vacuum, μ_0	1.257×10^{-6} H/m
Permittivity of vacuum, ε_0	8.854×10^{-12} F/m
Planck's constant, h	6.626×10^{-34} J/s
Velocity of light in vacuum, c	2.998×10^{8} m/s
Volume of perfect gas at standard temperature and pressure	22.41×10^{-3} m³/mol

Table A.2 Conversion of Units, General

Quantity	Imperial Unit	SI Unit
Angle, θ	1 rad	57.30°
Density, ρ	1 lb/ft³	16.03 kg/m³
Diffusion coefficient, D	1 cm²/s	1.0×10^{-4} m²/s
Energy, U	See below	
Force, F	1 kgf 1 lbf 1 dyne	9.807 N 4.448 N 1.0×10^{-5} N
Length, ℓ	1 ft 1 inch 1 Å	304.8 mm 25.40 mm 0.1 nm
Mass, M	1 tonne 1 short ton 1 long ton 1 lb mass	1000 kg 908 kg 1107 kg 0.454 kg
Power, P	See below	
Stress, σ	See below	
Specific heat, C_p	1 cal/gal °C Btu/lb °F	4.188 kJ/kg °C 4.187 kJ/kg °C
Stress intensity, K_{1c}	1 ksi$\sqrt{\text{in}}$	1.10 MN/m³/²
Surface energy γ	1 erg/cm²	1 mJ/m²

Table A.2 Conversion of Units, General—Cont'd

Quantity	Imperial Unit	SI Unit
Temperature, T	1 °F	0.556 °K
Thermal conductivity, λ	1 cal/sq cm °C 1 Btu/h ft °F	418.8 W/m °C 1.731 W/m °C
Volume, V	1 Imperial gall 1 US gall	4.546×10^{-3} m³ 3.785×10^{-3} m³
Viscosity, η	1 poise 1 lb/ft s	0.1 N s/m² 0.1517 N s/m²

Stress and pressure. The SI unit of stress and pressure is the N/m² or the Pascal (Pa), but from a materials point of view 1 Pa is very small. The levels of stress large enough to distort or deform materials are measured in mega Pascals (MPa). The table lists the conversion factors relating MPa to measures of stress used in the older cgs and metric systems (dyne/cm², kgf/mm²) by the imperial system (lb/in², ton/in²) and by atmospheric science (bar).

Table A.3 Conversion of Units – Stress and Pressure

To → From ↓	MPa	dyn/cm²	lb/in²	kgf/mm²	bar	long ton/ in²
			Multiply by			
MPa	1	10^7	1.45×10^2	0.102	10	6.48×10^{-2}
dyn/cm²	10^{-7}	1	1.45×10^{-5}	1.02×10^{-8}	10^{-6}	6.48×10^{-9}
lb/in²	6.89×10^{-3}	6.89×10^4	1	703×10^{-4}	6.89×10^{-2}	4.46×10^{-4}
kgf/mm²	9.81	9.81×10^7	1.42×10^3	1	98.1	63.5×10^{-2}
Bar	0.10	10^6	14.48	1.02×10^{-2}	1	6.48×10^{-3}
long ton/in²	15.44	1.54×10^8	2.24×10^3	1.54	1.54×10^2	1

Energy and power. The SI units of energy is the Joule (J), that of power is the Watt (W = 1 J/sec), or multiples of them like MJ or kW. If energy and power were always listed in these units, life would be simple, but they are not. First there are imperial units Btu, ft lbf and ft lbf/s. Then there are units of convenience: kWh for electric power, hp for mechanical power. There are the units of the oil industry: barrels (7.33 bbl = 1 tonne, 1000 kg) and toe (tonnes of oil equivalent – the weight of oil with the same energy content). Switching between these units is simply a case of multiplying by the conversion factors listed below.

Table A.4 Conversion of Units – Energy[a]

To → From ↓	MJ	kWh	kcal	Btu	ftlbf	toe
			Multiply by			
MJ	1	0.278	239	948	0.738×10^6	23.8×10^{-6}
kWh	3.6	1	860	3.41×10^3	2.66×10^6	85.7×10^{-6}
kcal	4.18×10^{-3}	1.16×10^{-3}	1	3.97	3.09×10^3	99.5×10^{-9}
Btu	1.06×10^{-3}	0.293×10^{-3}	0.252	1	0.778×10^3	25.2×10^{-9}
ftlbf	1.36×10^{-6}	0.378×10^{-6}	0.324×10^{-3}	1.29×10^{-3}	1	32.4×10^{-12}
toe	41.9×10^3	11.6×10^3	10×10^6	39.7×10^6	30.8×10^9	1

[a]MJ = megajoules; kWh = kilowatt hour; kcal = kilocalorie; Btu = British thermal unit; ftlbf = foot-pound force; toe = tonnes oil equivalent.

Table A.5 Conversion of Units – Power[a]

To From ↓	kW (kJ/s)	kcal/s	hp	ft lbf/s
		Multiply by		
kW (kJ/s)	1	4.18	1.34	735
kcal/s	0.239	1	0.321	176
hp	0.746	3.12	1	545
ft lbf/s	1.36×10^{-3}	5.68×10^{-3}	1.82×10^{-3}	1

[a]kW = kilowatt; kcal/s = kilocalories per second; hp = horse power; ft lb/s = foot-pounds/second.

A.2 MATERIALS

Elements. This table lists the concentration of elements in the earth's crust, that of a typical mined ore grade, and the present world production and the approximate price of commercial purity grades.

Table A.6 Abundance, Ore Grade, World Production and Approximate Price for the Elements

Element	Abundance in Earth's Crust PPM	Typical Ore Grade % Metal	World Production tonnes/Year	Price	
				₡/kg	$/kg
Aluminum	82e3	32	44e3	1.6	2.1
Antimony	0.32	13	190e3	4	5.3
Beryllium	2.2	0.45	140	830	1.1e3
Bismuth	20e-3	0.42	7.3e3	7.8	10
Cadmium	0.12	0.2	19e3	2.9	3.9
Calcium	45e3	40	2e3	8	11
Cerium	64	1.5	50e3	38	50
Chromium	120	32	23e6	7.8	10
Cobalt	24	0.55	72e3	20	26
Copper	58	2.6	16e6	6.4	8.5
Dysprosium	5.2	-	1.3e3	130	170
Erbium	3.5	-	500	320	430
Europium	1.9	-	590	3.5e3	4.7e3
Gadolinium,	6.4	-	2.3e3	120	160
Gallium	19	25e-3	78	460	610
Gold	1.8e-3	1.8e-3	2.4e3	36e3	48e3
Hafnium	4.2	16	90	96	130
Holmium	1.3	-	10	190	250
Indium	0.25	50e-3	600	630	840
Iridium	35e-6	-	3	9.8e3	13e3
Iron	51e3	-	2.3e9	0.39	0.52
Lanthanum	39	-	35e3	17	23
Lead	14	8	3.9e6	1.7	2.2
Lithium	20	0.6	18e3	49	66
Lutetium	0.53	-	10	9.2e3	12e3
Magnesium	23e3	0.15	570e3	2.5	3.3
Manganese	1e3	0.15	9.6e6	1.7	2.2
Mercury	58e-3	-	1.3e3	6.7	9

(Continued)

Table A.6 Abundance, Ore Grade, World Production and Approximate Price for the Elements—Cont'd

Element	Abundance in Earth's Crust PPM	Typical Ore Grade % Metal	World Production tonnes/Year	Price ₵/kg	Price $/kg
Molybdenum	1.3	-	200e3	33	43
Neodymium	41	-	22e3	40	53
Nickel	85	1	1.4e6	14	18
Niobium	18	1.7	62e3	190	250
Osmium	420e-6	-	50e-3	14e3	19e3
Palladium	15e-3	-	190	17e3	22e3
Platinum	5e-3	250e-6	180	39e3	52e3
Praseodymium	9.1	-	6.3e3	49	65
Rhenium	700e-6	0.5	52	800	1.1e3
Rhodium	370e-6	0.5	16	120e3	160e3
Samarium	6.9	-	2.7e3	100	140
Scandium	20	-	2e3	12e3	16e3
Selenium	50e-3	11	1.5e3	9.3	12
Silver	75e-3	55e-3	21e3	560	750
Tantalum	1.8	0.3	1.2e3	410	540
Tellurium	1e-3	4	200	86	120
Terbium	0.97	-	250	1.3e3	1.7e3
Thallium	0.61	75e-3	10	250	330
Thulium	0.47	-	50	4.9e3	6.5e3
Tin	2.1	2	310e3	17	23
Titanium	5.4e3	16	200e3	7.9	11
Tungsten	1	1.7	58e3	43	57
Uranium	1.9	0.2	40e3	120	150
Vanadium	170	1.2	54e3	970	1.3e3
Ytterbium	2.9	-	50	710	950
Yttrium	29	-	8.7e3	78	100

Granta Design Eco Design database 2014, www.grantadesign.com

Polymers. This table lists current world production and approximate price of common polymers.

Table A.7 World Production and Approximate Price of Polymers

Polymer	World Production tonnes/Yr	Price	
		¢/kg	$/kg
Acrylonitrile butadiene styrene (ABS)	5.6e6	2.1	2.7
Butyl rubber (IIR)	10e6	3.4	4.6
Cellulose polymers (CA)	-	3.2	4.3
Epoxies	120e3	2.6	3.5
Ethylene vinyl acetate (EVA)	-	1.6	2.1
Ionomer (I)	-	2.7	3.7
Phenolics	10e6	1.3	1.8
Polyamides (Nylons, PA)	3.7e6	3.6	4.8
Polycarbonate (PC)	-	3.7	4.9
Polychloroprene (Neoprene, CR)	-	4.5	6
Polyester	40e6	1.8	2.4
Polyetheretherketone (PEEK)		70	93
Polyethylene (PE)	68e6	1.4	1.8
Polyethylene terephthalate (PET)	9.1e6	1.6	2.2
Polyhydroxyalkanoates (PHA, PHB)	-	4.9	6.5
Polyisoprene rubber (IIR)	-	2.9	3.9
Polylactide (PLA)	-	1.8	2.4
Polymethyl methacrylate (Acrylic, PMMA)	-	2.2	2.9
Polyoxymethylene (Acetal, POM)	-	2.5	3.3
Polypropylene (PP)	43e6	1.3	1.7

(Continued)

Table A.7 World Production and Approximate Price of Polymers—Cont'd

Polymer	World Production tonnes/Yr	Price	
		₡/kg	$/kg
Polystyrene (PS)	12e6	2.7	3.6
Polytetrafluoroethylene (Teflon, PTFE)	-	12	16
Polyurethane	-	4.6	6.1
Polyurethane (tpPUR)	-	4.6	6.1
Polyvinylchloride (tpPVC)	-	1.1	1.5
Silicone elastomers (SI, Q)	-	8.6	11
Starch-based thermoplastics (TPS)	-	4.9	6.5

Granta Design Eco Design database 2014, www.grantadesign.com

Other materials. This table lists current world production and approximate price of wood, paper, glass, building materials and composites

Table A.8 World Production and Approximate Price of Diverse Materials

Material	World Production tonnes/Year	Price	
		₡/kg	$/kg
Alumina	1.2e6	17	22
Bamboo	1.4e9	1.2	1.6
Brick	50e6	0.76	1
Cement	750e6	0.082	0.11
CFRP	28e3	30	39
Concrete	15e9	0.037	0.049

Table A.8 World Production and Approximate Price of Diverse Materials   Cont'd

Material	World Production tonnes/Year	Price ¢/kg	Price $/kg
Paper and cardboard	360e6	0.82	1.1
Soda-lime glass	81e6	1.1	1.5
Softwood	960e6	0.71	0.95

Granta Design Eco Design database 2014, www.grantadesign.com

A.3   ENERGY

When comparing energy use, it is normal practice to convert all energies to kg oil equivalent (kg OE). The use of fuels – even nuclear fuels – releases emissions, among them CO_2 and other oxides of carbon, nitrogen and sulfur contribute to global warming. The global warming potential (GWP) is the CO_2 equivalent ($CO_{2,eq}$) of these emissions.

Table A.9 The Energy Intensity of Fuels and Their Carbon Footprints

Fuel Type	kg OE[a]	MJ/liter	MJ/kg	$CO_{2,eq}$ kg/liter	$CO_{2,eq}$ kg/kg	$CO_{2,eq}$ kg/MJ
Coal, lignite	0.45	-	18–22	-	1.6	0.080
Coal, anthracite	0.72	-	30–34	-	2.9	0.088
Crude oil	1.0	38	44	3.1	3.0	0.070
Diesel	1.0	38	44	3.1	3.2	0.071
Gasoline	1.05	35	45	2.9	2.89	0.065
Kerosene	1.0	35	46	3.0	3.0	0.068
Ethanol	0.71	23	31	2.8	2.6	0.083
LNG	1.2	25	55	3.03	3.03	0.055
Hydrogen	2.7	8.5	120	0	0	0

[a]kilograms oil equivalent (the kg of oil with the same energy content).

Table A.10 Electricity Generation, Energy Mix, MJ Oil per kWh and $CO_{2,eq}$ per kWh[a]

Country	Fossil Fuel %	Nuclear %	Renewables %	Efficiency (a) %	MJ_{oe}(b) per kWh(d)	$CO_{2,eq}$(c), kg per kWh(d)
Australia	92	0	8	33	10.0	0.71
China	83	2	15	32	9.3	0.66
France	10	78	12	40	0.9	0.06
India	81	2.5	16.5	27	10.8	0.77
Japan	61	27	12	41	5.4	0.38
Norway	1	0	99	-	0	0
UK	75	19	6	40	6.6	0.47
USA	71	19	10	36	7.1	0.50
OEDC (Europe)	62	22	16	39	5.7	0.41
World average	67	14	19	36	6.7	0.48

(a) Conversion efficiency of fossil fuel to electricity; (b) MJoe of fossil fuel (oil equivalent) used in energy mix per kWh of delivered electricity from all sources; (c) CO_2 equivalent release per kWh of delivered electricity from all sources; (d) 1 kWh is 3.6 MJelectric.
[a]Data from IEA (2008).

Table A.11 Alternative Measures for Oil and Gas (Coal is Measured in tonnes or Short tons)

Crude oil	Natural gas
1 bbl = 35 Imperial gallons =42 US gallons =159 liters ≈135 kg	1 billion m^3 $(10^9 m^3) = 35.5 \times 10^9 ft^3$ $=6.29 \times 10^6$ boe* $=0.9 \times 10^6$ toe** $=0.73 \times 10^6$ tonnes LNG**

*boe = Barrel of oil equivalent; **toe = Tonne of oil equivalent; ***LNG = iquid natural gas.

Table A.12 Average Approximate Gobal Resource Intensities and Capacity Factors for Power Generating Systems

Power System	Capital Intensity (k$/kW$_p$)[a]	Area Intensity (m²/kW$_p$)	Material Intensity (kg/kW$_p$)	Construction Energy Intensity (MJ/kW$_p$)	Construction Carbon Intensity (kg/kW$_p$)	Capacity Factor (%)
Conventional gas	0.6–1.5	1–4	605–1080	1,730–2,710	100–200	75–85
Conventional, coal	2.5–4.5	1.5–3.5	700–1600	3,580–9,570	100–700	75–85
Phosphoric acid fuel cell	3–4.5	0.1–0.5	80–120	5,000–10,000	600–1000	>95
Solid oxide fuel cell	7–8	0.3–1	50–100	2,000–6,000	200–400	>95
Nuclear - fission	3.5–6.4	1–3	170–625	2,000–4,300	105–330	75–95
Wind, land-based	1.0–2.4	150–400	500–2,000	3,500–6,000	240–600	17–25
Wind, off-shore	1.6–3	100–300	300–900	5,000–10,000	480–1,000	30–40
Solar PV, single crystal	4–12	30–70	800–1,700	30,000–60,000	2,000–4,000	8–12[b]
Solar PV, poly-silicon	3–6[b]	50–80[b]	1,000–2,000	20,000–40,000	1,500–3,000	8–12[b]
Solar PV, thin-film	2–5	50–100	1,500–3,000	10,000–20,000	550–1,000	8–12[b]
Solar thermal	3.9–8	20–100	650–3,500	19,000–40,000	1,500–3,500	20–35[c]
Hydro-earth dam	1–5	200–600	15,000–100,000	7260–15,000	630–1,200	45–65

(Continued)

Table A.12 Average Approximate Gobal Resource Intensities and Capacity Factors for Power Generating Systems—Cont'd

Power System	Capital Intensity (k$/kW$_p$)[a]	Area Intensity (m^2/kW$_p$)	Material Intensity (kg/kW$_p$)	Construction Energy Intensity (MJ/kW$_p$)	Construction Carbon Intensity (kg/kW$_p$)	Capacity Factor (%)
Hydro- steel reinforced	1–5	120–500	8,000–40,000	30,000–66,000	1,000–4,000	50–70
Wave	1.2–4.4	42–100	1,000–2,000	22,950–31,540	1670–2070	25–40
Tidal-current	10–15	150–200	350–650	12,000–18,000	800-1130	35–50
Tidal-barrage	1.6–2.5	200–300	5,000–50,000	30,000–45,000	2,400–3,520	20–30
Geothermal-shallow	1.15–2	1–3	61–500	7,000–13,500	160–250	75–95
Geothermal-deep	2–3.9	1–3	400–1200	20,000–40,700	1,700–3,900	75–95
Biomass-dedicated	2.3–3.6	10,000–33,000	500–922	5,000–19,800	600–1800	75–95

Data from "Materials and the Environment" by M.F. Ashby, Butterworth Heinemann, (2013).

[a]kW$_p$ means peak (or rated) electrical power output. The actual output is the peak multiplied by the capacity factor.

[b]Estimated capacity factor for PV in the UK and equivalent latitudes in Europe. The capacity factor for PV in central Australian, Sahara or Mojave deserts could be four times greater.

[c]Typical capacity factor for solar thermal built in a suitable location, such as Spain, North Africa, Australia, or Southern USA.

Table A.13 Performance Metrics of Energy Storage Systems[a]

Storage System	Specific Energy (MJ/kg)	Energy Density (MJ/m³)	Operating Cost ($/MJ)	Specific Power (W/kg)	Efficiency (%)	Cycle Life (#-cycles)[b]
Conventional fuels	20–50	15,000–72,000	-	-	18–50	1
Pumped hydro	0.002–0.005	2–5	0.0006–0.0014	0.02–0.3	70–80	400,000
Compressed air energy storage	0.36	25	0.0001–0.0019	8	65–70	15,000
Springs	0.00014–0.00033	0.6–1.1	-	-	98–99.9	Depends on loading[1]
Flywheels	0.002–0.025	2.1	0.0008–0.0017	100–10,000	75–85	150,000
Thermal storage	0.032–0.036	160	0.0008–0.0019	1.5–1.7	72–85	10,000–15,000
Li-ion batteries	0.32–0.68	720–1,400	0.0019–0.0047	250–340	80–90	300–2,000
Sodium–sulfur batteries	0.2–0.7	140–540	0.0039	10–15	75–83	3600–4700
Lead-acid batteries	0.07–0.18	200–430	0.0008–0.0028	4–180	70–90	200–1500
Nickel cadmium batteries	0.08–0.23	72–310	0.0008–0.0055	30–150	60–85	800–1200
Nickel-metal hydride batteries	0.1–0.43	190–1300	0.0006–0.0028	4–140	65–85	300–1000
Vanadium flow batteries	0.07–0.13	110–170	0.0039	2.2–3.1	71–88	10000–16000
Hydrogen fuel cells	0.02–0.8	1–70	0.0006–0.0041	5–20	27–35	5,000–10,000
Super capacitors	0.010–0.020	10–25	0.0008–0.0019	1,000–20,000	90–95	1,000,000
Superconducting magnetic energy storage	0.001–0.02	2–10	0.0003–0.0083	500–20,000	85–92	50,000–200,000

[a]Data from "Materials and the Environment" by M.F. Ashby, Butterworth Heinemann, (2013).
[b]The cycle life of a spring depends on loading – for low alloy steels, for example, an increase in loading of 50% can reduce the lifetime by six orders of magnitude.

Table A.14 The Approximate Energy and Carbon Footprint of Transport[a]

Transport Type and Fuel	Energy (MJ/tonne.km[b])	Carbon footprint (kg$CO_{2,eq}$/tonne.km[b])
Ocean shipping – Diesel	0.16	0.015
Coastal shipping – Diesel	0.27	0.019
Barge – Diesel	0.36	0.028
Rail – Diesel	0.25	0.019
Articulated heavy goods vehicle (up to 55 tonne) – Diesel	0.71	0.05
40 tonne truck – Diesel	0.82	0.06
32 tonne truck – Diesel	0.94	0.067
14 tonne truck – Diesel	1.5	0.11
Light goods vehicle – Diesel	2.5	0.18
Family car – Diesel	1.4–2.0	0.1–0.14
Family car – Gasoline	2.2–3.0	0.14–0.19
Family car – LPG	3.9	0.18
Family car – Hybrid gasoline-electric	1.55	0.10
Super sports car and SUV – Gasoline	4.8	0.31
Long haul aircraft – Kerosene	6.5	0.45
Short haul aircraft – Kerosene	11–15	0.76
Helicopter (Eurocopter AS 350) – Kerosene	55	3.30

[a]Data sources are listed under Further Reading.
[b]1 tonne mile = 1.46 tonne.km

Table A.15 Some Approximate Efficiency Factors for Energy Conversion

Energy Conversion Path	Efficiency, Direct Conversion (%)	Efficiency Relative to Oil Equivalence (%)	Associated Carbon (kg $CO_{2,eq}$ per Useful MJ)
Gas to electric	37–40	37–40	0.18
Oil to electric	36–38	36–38	0.2
Coal to electric	33–35	33–35	0.22
Biomass to electric	23–26	23–26	0
Hydro to electrical	75–85	75–85	0
Nuclear to electric	32–34	32–34	0
Fossil fuel to heat, enclosed system	100	100	0.07
Fossil fuel to heat, vented system	65–75	65–75	0.10
Biomass to heat, vented system	33–35	33–35	0
Oil to mechanical, diesel engine	20–22	19–21	0.17
Oil to mechanical, petrol engine	13–15	12–14	0.15
Oil to mechanical, kerosene gas turbine	27–29	25–27	0.15
Electric to thermal	100	37–40	0.20
Electric to mechanical (electric motors)	80–93	30–32	0.23
Electric to chemical (lead-acid battery)	80–85	29–32	0.24
Electric to chemical (advanced battery)	85–90	31–33	0.23

(Continued)

Table A.15 Some Approximate Efficiency Factors for Energy Conversion—Cont'd

Energy Conversion Path	Efficiency, Direct Conversion (%)	Efficiency Relative to Oil Equivalence (%)	Associated Carbon (kg $CO_{2,eq}$ per Useful MJ)
Electric to electromagnetic radiation (incandescent lamp)	2–3	0.6–1	4.5
Electric to electromagnetic radiation (CFL)	8–11	2.8–4	0.7
Electric to electromagnetic radiation (LED)	9–15	3–5	0.5
Light to electric (solar cell)	10–22	-	0

Table A.16 Approximate Energy and $CO_{2,eq}$ Rating of Cars as a Function of Their Mass

Fuel Type	Energy per km – Mass (H_{km} in MJ/km, m in kg)	CO_2 per km – Mass ($CO_{2/km}$ in g/km, m in kg)	dH_{km}/dm MJ/km kg ($m = 1000$ kg)
Petrol power	$H_{km} \approx 3.7 \times 10^{-3}\, m^{0.93}$	$CO_2 \approx 0.25\, m^{0.93}$	2.1×10^{-3}
Diesel power	$H_{km} \approx 2.8 \times 10^{-3}\, m^{0.93}$	$CO_2 \approx 0.21\, m^{0.93}$	1.6×10^{-3}
LPG power	$H_{km} \approx 3.7 \times 10^{-3}\, m^{0.93}$	$CO_2 \approx 0.17\, m^{0.93}$	2.2×10^{-3}
Hybrid power	$H_{km} \approx 2.3 \times 10^{-3}\, m^{0.93}$	$CO_2 \approx 0.16\, m^{0.93}$	1.3×10^{-3}

A.4 ENVIRONMENT

Elements. This tables list the embodied energies, carbon footprint and approximate water usage per unit mass for the elements.

Table A.17 Approximate Environmental Data for the Elements

Element	Embodied Energy MJ/kg	$CO_{2,eq}$ (GWP) kg/kg	Water Usage l/kg
Aluminum	210	13	1.2e3
Antimony	140	13	3.4e3
Beryllium	6.4e3	400	360
Bismuth	150	9.3	2.9e3
Cadmium	13	0.68	820
Calcium	150	9.5	-
Cerium	510	33	55
Chromium	300	16	470
Cobalt	130	8.3	580
Copper	60	3.7	310
Dysprosium	1.4e3	90	55
Erbium	2.9e3	190	55
Europium	20e3	1.3e3	55
Gadolinium	1.3e3	86	55
Gallium	3.1e3	210	3.8e3
Gold	250e3	15e3	-
Hafnium	1.1e3	69	8.1e3
Holmium	1.9e3	120	55
Indium	2.9e3	180	1.3e3
Iridium	45e3	2.9e3	-
Iron	45	3	50
Lanthanum	290	19	55
Lead	27	1.9	340
Lithium	590	28	470

(Continued)

Table A.17 Approximate Environmental Data for the Elements—Cont'd

Element	Embodied Energy MJ/kg	$CO_{2,eq}$ (GWP) kg/kg	Water Usage l/kg
Lutetium	48e3	3e3	55
Magnesium	320	36	980
Manganese	65	4	240
Mercury	130	8.8	470
Molybdenum	380	34	360
Neodymium	85	5.5	55
Nickel	170	11	230
Niobium	1.7e3	110	150
Osmium	62e3	4e3	-
Palladium	160e3	8.5e3	-
Platinum	270e3	15e3	-
Praseodymium	650	42	55
Rhenium	12e3	770	3.8e3
Rhodium	560e3	30e3	-
Samarium	1.2e3	77	55
Scandium	59e3	3.8e3	55
Selenium	72	4.5	26
Silver	1.5e3	100	76e3
Tantalum	4.3e3	260	650
Tellurium	160	7.5	26
Terbium	9e3	580	55
Thallium	2.4e3	150	1.8e3
Thulium	26e3	1.7e3	55
Tin	230	13	11e3
Titanium	580	39	110
Tungsten	540	34	150
Uranium	1.3e3	81	3.5e3
Vanadium	3.7e3	250	650
Ytterbium	5.6e3	360	55
Yttrium	1.5e3	93	55

Granta Design Sustainability database 2014, www.grantadesign.com

Polymers. This tables list the embodied energies, carbon footprint and approximate water usage per unit mass for the polymers.

Table A.18 Approximate Environmental Data for Polymers

Polymer	Embodied Energy MJ/kg	$CO_{2,eq}$ (GWP) kg/kg	Water Usage l/kg
Acrylonitrile butadiene styrene (ABS)	95	3.8	180
Butyl rubber (IIR)	120	6.6	110
Cellulose polymers (CA)	90	3.8	240
Epoxies	130	7.2	28
Ethylene vinyl acetate (EVA)	79	2.1	2.8
Ionomer (I)	110	4.2	280
Phenolics	79	3.6	52
Polyamides (Nylons, PA)	120	8	180
Polycarbonate (PC)	110	6	170
Polychloroprene (Neoprene, CR)	64	1.7	220
Polyester	40e6	3	200
Polyetheretherketone (PEEK)	300	23	920
Polyethylene (PE)	81	2.8	58
Polyethylene terephthalate (PET)	85	4	130
Polyhydroxyalkanoates (PHA, PHB)	85	4.4	170
Polyisoprene rubber (IIR)	100	5.4	150
Polylactide (PLA)	52	3.6	69
Polymethyl methacrylate (Acrylic, PMMA)	110	6.8	76
Polyoxymethylene (Acetal, POM)	90	4	240
Polypropylene (PP)	80	3.1	39

(Continued)

Table A.18 Approximate Environmental Data for Polymers—Cont'd

Polymer	Embodied Energy MJ/kg	$CO_{2,eq}$ (GWP) kg/kg	Water Usage l/kg
Polystyrene (PS)	97	3.8	140
Polytetrafluoroethylene (Teflon, PTFE)	110	6	460
Polyurethane	87	3.7	98
Polyurethane (tpPUR)	87	3.7	98
Polyvinylchloride (tpPVC)	58	2.5	210
Silicone elastomers (SI, Q)	120	7.9	330
Starch-based thermoplastics (TPS)	25	1.1	170

Granta Design Sustainability database 2014, www.grantadesign.com

Other materials. This table lists the embodied energies, carbon footprint and approximate water usage per unit mass wood, paper, glass, building materials and composites

Table A.19 Approximate Environmental Data for Diverse Materials

Material	Embodied Energy MJ/kg	$CO_{2,eq}$ (GWP) kg/kg	Water Usage l/kg
Alumina	52	2.8	57
Bamboo	5	0.31	700
Brick	3.3	0.22	5.5
Cement	5.7	0.95	37
CFRP	480	35	1.4e3
Concrete	1.1	95e-3	3.4
Paper and cardboard	51	1.2	1.7e3
Soda-lime glass	11	0.76	14
Softwood	9.2	0.38	700

Granta Design Sustainability database 2014, www.grantadesign.com

A.5 ECONOMICS (2011 DATA)

Material prices have already been listed in Tables A.6, A.7 and A.8. Energy prices fluctuate[1]. All, ultimately, are tied to the price of oil, which in July 2014 stood at $103/bbl.

Table A.20 GDP per Capita and Price of Energy and Water

Nation	GDP per Capita (US$)	Price of Gasoline (US$/litre)	Price of Electricity (US$/kWh)	Price of Water (US$/m³)
Algeria (ALG)	1.7e3	0.29	-	-
Argentina (ARG)	3.3e3	1.5	0.058	-
Australia (AUS)	20e3	1.4	0.32	2.4
Austria (AUT)	24e3	1.8	0.25	-
Bangladesh (BGD)	290	1.2	-	-
Belgium (BEL)	22e3	2.1	0.29	4.0
Botswana (BOT)	3.3e3	1.2	-	-
Brazil (BRA)	3.2e3	1.4	0.34	-
Burundi (BDI)	58	1.5	-	-
Cambodia (CAM)	290	1.4	-	-
Canada (CAN)	20e3	1.3	0.087	1.6
Chile (CHI)	3.8e3	1.6	0.23	-
Congo, D.R. (COD)	76	1.5	-	1.4
Cuba (CUB)	2.2e3	1.4	-	-
Denmark (DEN)	29e3	2	0.38	6.7
Egypt (EGY)	660	0.45	-	-

(Continued)

---

[1] https://www.gov.uk/government/statistical-data-sets/annual-domestic-energy-price-statistics

Table A.20 GDP per Capita and Price of Energy and Water—Cont'd

Nation	GDP per Capita (US$)	Price of Gasoline (US$/litre)	Price of Electricity (US$/kWh)	Price of Water (US$/m³)
Ethiopia (ETH)	110	1.1	-	-
Fiji Islands (FIJ)	1.4e3	1.4	-	-
Finland (FIN)	23e3	2.1	0.2	4.4
France (FRA)	21e3	1.9	0.18	3.7
Germany (GER)	22e3	2	0.33	-
Greece (GRE)	13e3	2.1	0.16	-
Guatemala (GUA)	1.7e3	1.1	-	-
India (IND)	440	1.3	0.098	-
Indonesia (INA)	700	0.47	0.088	-
Ireland (IRL)	31e3	2	0.26	-
Israel (ISR)	14e3	-	0.18	-
Japan (JPN)	24e3	2	0.22	1.9
Kenya (KEN)	300	1.4	-	-
Libya (LBA)	3.5e3	0.12	-	-
Luxembourg (LUX)	45e3	1.6	0.21	-
Mexico (MEX)	4.2e3	0.86	0.19	0.49
Morocco (MAR)	1.1e3	1.4	-	-
New Zealand (NZL)	16e3	1.8	0.19	2.0
Norway (NOR)	36e3	2.5	-	-
P.R. of China (CHN)	1.2e3	1.4	0.091	-

Table A.20 GDP per Capita and Price of Energy and Water—Cont'd

Nation	GDP per Capita (US$)	Price of Gasoline (US$/litre)	Price of Electricity (US$/kWh)	Price of Water (US$/m³)
Pakistan (PAK)	470	1.1	0.055	-
Papua New Guinea (PNG)	450	0.94	0.29	-
Peru (PER)	1.7e3	1.6	0.1	-
Russia (RUS)	3.2e3	0.99	0.048	-
Rwanda (RWA)	120	1.7	-	-
Singapore (SIN)	17e3	1.7	0.22	-
Somalia (SOM)	170	1.1	-	-
South Africa (RSA)	2.8e3	1.4	0.11	-
South Korea (KOR)	11e3	1.8	-	0.77
Spain (ESP)	17e3	1.8	0.24	1.9
Sweden (SWE)	26e3	2.1	0.25	3.6
Turkey (TUR)	3.2e3	2.5	-	-
Uganda (UGA)	180	1.4	-	-
United Arab Emirates (UAE)	15e3	0.47	0.076	-
United Kingdom (GBR)	24e3	2.2	0.19	4.8
United States of America (USA)	27e3	0.97	0.17	-
Zimbabwe (ZIM)	160	1.5	-	-

Granta Design Sustainability database 2014, www.grantadesign.com

A.6 SOCIETY

This table lists societal metrics of education, governance and health.

Table A.21 Literacy, Freedom of Speech, Corruption Control and Life Expectancy

Nation	Literacy (%)	Press Freedom (0 to 100)	Corruption Control (0 to 100)	Life Expectancy (Years)
Afghanistan (AFG)	18	45	1	49
Algeria (ALG)	73	57	38	73
Argentina (ARG)	98	84	40	76
Australia (AUS)	99	91	96	82
Austria (AUT)	98	99	92	81
Bangladesh (BGD)	55	56	16	69
Belgium (BEL)	99	95	90	80
Botswana (BOT)	83	86	80	53
Brazil (BRA)	90	70	60	74
Brunei (BRU)	95	56	78	78
Burundi (BDI)	66	55	12	50
Cambodia (CAM)	78	57	8	63
Canada (CAN)	99	97	97	81
Chile (CHI)	99	74	91	79
Congo, D.R.(COD)	67	49	3	48
Cuba (CUB)	100	28	72	79
Denmark (DEN)	99	97	100	79
Egypt (EGY)	66	29	34	73

Table A.21 Literacy, Freedom of Speech, Corruption Control and Life Expectancy—Cont'd

Nation	Literacy (%)	Press Freedom (0 to 100)	Corruption Control (0 to 100)	Life Expectancy (Years)
Ethiopia (ETH)	36	56	28	59
Fiji Islands (FIJ)	94	57	19	69
Finland (FIN)	100	100	98	80
France (FRA)	99	87	89	82
Germany (GER)	99	95	93	80
Greece (GRE)	97	78	56	80
India (IND)	63	55	36	65
Indonesia (INA)	92	49	27	69
Ireland (IRL)	99	96	93	81
Israel (ISR)	92	73	72	82
Japan (JPN)	99	94	92	83
Kenya (KEN)	87	74	19	57
Libya (LBA)	90	42	6	75
Luxembourg (LUX)	100	98	95	80
Mexico (MEX)	93	46	44	77
Morocco (MAR)	56	52	53	72
New Zealand (NZL)	99	97	100	81
Norway (NOR)	100	100	97	81
P.R. of China (CHN)	94	3.9	33	74
Pakistan (PAK)	54	44	12	65

(Continued)

Table A.21 Literacy, Freedom of Speech, Corruption Control and Life Expectancy—Cont'd

Nation	Literacy (%)	Press Freedom (0 to 100)	Corruption Control (0 to 100)	Life Expectancy (Years)
Papua New Guinea (PNG)	60	88	10	63
Peru (PER)	90	60	50	74
Russia (RUS)	100	50	13	69
Rwanda (RWA)	70	40	71	55
Singapore (SIN)	95	53	99	81
South Africa (RSA)	89	86	61	53
South Korea (KOR)	98	85	69	81
Spain (ESP)	98	87	81	81
Sweden (SWE)	99	97	99	81
Turkey (TUR)	89	47	58	74
Uganda (UGA)	75	51	21	54
United Arab Emirates (UAE)	90	64	80	77
United Kingdom (GBR)	99	92	90	80
United States of America (USA)	99	84	86	79
Zimbabwe (ZIM)	91	57	2	51

Granta Design Sustainability database 2014, www.grantadesign.com

Further Reading

Ashby MF: *Materials and the environment*, 2nd ed., Oxford, 2013, Butterworth Heinemann. ISBN 978-0-12-385971-6.

Carbon Trust: *Carbon footprint in the supply chain*. www.carbontrust.co.uk, 2007.

CES EduPack Eco-materials database, Granta Design, Cambridge: www.grantad esign.com, 2014.

Department of Energy and Climate Change: www.statistics.gov.uk/hub/busin ess-energy/energy/energy-prices, 2011.

Electricity information: IEA publications, 2008. ISBN 978-9264-04252-0. *(One of a series of IEA statistical publications, this one giving every statistic you ever wanted to know and plenty you don't about electricity.)*

International Chamber of Shipping: In *International shipping federation, annual review*, 2005. www.marisec.org/annualreview/annualreview.pdf. (Energy per ton-mile for shipping).

Jancovici J-M: http://www.manicore.com, 2007 (A mine of useful information about energy).

MacKay DJC: *Sustainable energy – without the hot air*, Cambridge, UK, 2008, Department of Physics, Cambridge University. www.withouthotair.com/ (Helpful assemble of useful and quirky numbers relevant to energy generation and use.)

Network Rail: www.networkrail.co.uk/freight/, 2007 (Energy per tonne.km for rail transport).

Nielsen R: *The little green handbook*, Carlton North, Victoria, Australia, 2005, Scribe Publications Pty Ltd. ISBN 1-9207-6930-7 (Well researched tables of energy information).

Reporters Without Borders: http://en.rsf.org/, 2013.

Shell Petroleum: *How the energy industry works*, Towchester, UK, 2007, Silverstone Communications Ltd. ISBN 978-0-9555409-0-5 (Both BP and Shell publish annual compilations of energy statistics).

Transport Watch UK: www.Transwatch.co.uk/Transport-fact-sheet-5, 2007 (More on energy for transport).

UCDP Conflict Encyclopedia: www.ucdp.uu.se/gpdatabase/search.php, 2013.

USGS: *Minerals information: mineral commodity summaries*, (The bible of resource data) http://minerals.usgs.gov/minerals/pubs/commodity/, 2007.

World Bank Worldwide Governance Indicators: www.govindicators.org, 2011.

Index

Note: Page numbers followed by "b", "f" and "t" indicate boxes, figures and tables respectively

A

Abundance risk
  anticipating, 257
  for earth elements, 95f
  in materials supply-chain,
    94–95
Accountancy, 30–31
Active stock, 220–221, 221f
Affluence
  consumption relating to, 17–19
  global, 253
  population relating to, 17–19
  summary of, 20
Agriculture, 129, 131
Aircrafts, 13, 15f
Algae, 131–132
Alternatives. *See also* Reflection
  for oil and gas, 286t
  power generating sources, 249f
Altschuller, Genrich, 76–77
Analogy
  case-based reasoning, inductive
    reasoning, and, 75–76
  TRIZ, 76–77
AngloAmerican case study,
    106b–107b
Animal feedstock, 129, 131
Appliances, 225
  washing machine eco-audit, 248f
Area-intensity, 144f
Articulation assessment tools,
    prompts, and checklists
  articulation fact-finding, 60–66
  creativity aids, 72–81
  exercises, 82–84
  informed synthesis, 66–70
  introduction to, 56
  prime objective clarification,
    56–57
  reflection, 71
  stakeholder analysis, 57–60
  summary of, 71–72
Articulations
  assessing, 34
  capitals relating to, 39

examples of, 33–36, 57t
generalizations about, 34
gone wrong, 113–115
human capital relating to, 68b
major sectors, 35, 36f
manufactured capital relating
  to, 68b
motivations of, 33–34
natural capital relating to, 68b
prime objective, 33–34, 56–57
stakeholders in, 34, 43–45,
  57–60, 68b
of sustainable development,
  33–36
Assessment rubric, 267–268, 269t
Automobile associations, 155
Automobiles. *See* Vehicles

B

Backcasting, 76
Bamboo
  belt, 198f
  global production of, 198, 204t
  human and social capital
    relating to, 208
  manufactured and financial
    capital relating to, 208
  natural capital relating to, 208
  producing nations and stock,
    204t
Bamboo flooring
  background on, 199–200
  carbon footprint of, 204t, 206
  concerns, 202b
  economics relating to, 206
  energy for, 204–205, 205f
  on environment, 205
  fact-finding on, 202–207, 203f
  introduction to, 198–200
  legislation on, 206, 207t
  long-term analysis, 208–210
  manufacturing, 199
  materials in, 202–203
  prime objective of, 200
  projects relating to, 210

properties of, 198–199, 199t
reflection on, 208–210
short-term analysis, 208
society relating to, 206–207
stakeholders, 200–202, 201f
synthesis, with three capitals,
  207–208, 209t
Batteries
  for electric cars, 155,
    157–158, 160
  rechargeable, 93–94
Better business models,
    231–233
Bicycles, 52, 166
Bill of materials (BOMs), 60–61
Bio-based Society, 52, 119
Biocomposites, 132
  in vehicles, 274
Biofuels, 33–34, 132–133
Biogenetic carbon dioxide, 205
Bioplastics, 59, 133
Biopolymers
  advocates of, 33–34
  animal feedstock relating to,
    129, 131
  background on, 119–120
  Bio-based Society on, 52, 119
  carbon footprint, 125f, 129
  commercial, 119t
  concerns, 122b
  corn in, 120
  economics and, 126–127, 131
  energy in, 123–124, 125f
  environmental properties,
    125–126, 125f
  fact-finding on, 122–127, 123f
  financial capital relating to,
    129–130
  fossil hydrocarbons in, 129
  global production of, 119–120,
    131
  human capital relating to, 130
  introduction to, 118–120
  legislation on, 126
  long-term analysis, 131–132

Biopolymers (*Continued*)
  manufactured capital relating to, 129–130
  material attributes, 123
  natural capital relating to, 129
  new industries for, 130
  oil-based, 119t
  personal satisfaction from, 130
  polymer pollution relating to, 129–130
  prime objective of, 120, 130
  projects related to, 132–133
  public information on, 130
  reflection on, 130–132
  short-term analysis, 130–131
  social capital relating to, 130
  stakeholders in, 120–122
  synthesis of, 127–130, 128t
  three capitals relating to, 127–130, 128t
  water in, 125f
  Young's modulus of, 123, 124f
BOMs. *See* Bill of materials
Bottom lines, 30–31
Brainstorming, 73–74
Brundtland Report, 4, 6f
Business. *See also* Corporate responsibility; Corporate sustainability
  community, 6
  sustainable, 30–31

C

CAFE rules. *See* Corporate Average Fuel Economy
Candela (*cd*), 168–169
Capacity factors, 183
  for power systems, 287t–288t
Capital. *See also* Three capitals; *specific capital*
  articulations relating to, 39
  base value, 48
  economy relating to, 32, 70
  environment relating to, 70
  fairness relating to, 70
  key features of, 67t
  net comprehensive, 32
  ranking, 48
  3Ps of, 28
Carbon cycle, 214–215, 214f
  burdens on, 215f
Carbon dioxide
  biogenetic, 205
  processing, 205

recovery, 206
  vehicle carbon dioxide ratings, 292t
Carbon emissions, 52
Carbon footprint
  of alternative power generating sources, 249f
  of bamboo floors, 204t, 206
  biopolymers, 125f, 129
  from coal, 143f
  of electric cars, 158–159
  of electrical power generation, 248–249
  of fuel, 62t, 285t
  from gas, 143f
  of information, 234b
  of solar PV, 186–187, 188f, 189
  of transport systems, 290t
Carbon taxes, 33–34
Carbon-offsetting, 114b–115b
Cars. *See* Vehicles
Case studies
  AngloAmerican, 106b–107b
  articulations gone wrong, 113–115
  on electric cars, 151–166
  exercises on, 116
  fact-finding in, 112
  introduction and synopsis, 111–112
  Jaguar Land Rover, 107b–108b
  prime objective of, 112
  reflection of, 113
  on SR, 105–108
  stakeholder in, 112
  structure of, 112–113
  summary of, 115–116
  synthesis, 112–113
Case-based reasoning, 75–76
Caterpillar Inc., 227b–228b
cd. See Candela
Cell phones
  critical materials in, 245, 245t
  Mazuma Mobile, 226b
  rare earth elements in, 272–273
  smart phone lifespan, 226b
  solar chargers for, 53
CES EduPack Sustainability database
  description of, 242–243
  structure and content, 243f
  using, 242–244
CFLs. *See* Lighting
Charts. *See* Data, charts, and databases

Chemical backgrounds, 93–94
China, 22–23
Chromium, 93–94
Circular materials economy
  active stock in, 220–221, 221f
  background and description of, 219–222
  better business models in, 231–233
  circularity metrics, 230–231, 230–231, 231f
  consumption in, 221–222
  contributions to, 235f
  creation of, 222–234
  design, for reuse, repair, and recycling, 225–230
  design improvements, 223–231, 224f
  ecological metaphor of, 213–217
  exercises on, 236–237
  improved materials technology, 222–223, 224f
  introduction to, 212–213
  recovery obstacles, 229–230, 229f
  regulation, 233
  replacing materials, by information, 233
  scale of, 217–219
  social and lifestyle changes, 233–234
  summary of, 234–236
Circularity metrics, 230–231, 230–231, 231f
Club of Rome, 3–4
Coal, 143f
Cobalt
  in electric cars, 166
  price volatility of, 88, 89f
  sourcing, 99
Coffins, 52–53
Commercial biopolymers, 119t
Commodity polymers, 10
Communities, 140, 144. *See also* Society
Community's CR Index, 104
Compact fluorescent lights (CFLs). *See* Lighting
Company performance, 104
Complex systems
  dealing with, 40–41
  incommensurate quantities in, 40
  layering of, 39f, 40, 41f

Conflict
  material risk, 257
  on-going, 92f
  risk, in materials supply-chain,
    91–92
Consumption. *See also* Global
    consumption
  affluence relating to, 17–19
  in circular materials economy,
    221–222
  production compared to, 222
  of resources, 17–19
Copper, 88, 89f
Corn
  in biopolymers, 120
  ethanol, 120
Corporate Average Fuel Economy
    (CAFE), 252
Corporate responsibility
  changing expectations of, 95–96
  Dodd-Frank act on, 95–96
  exercises on, 108–109
  in materials supply-chain, 95–96
  summary of, 108
Corporate social responsibility
    (CSR), 102–105
  defining, 102
  expectations for, 102
Corporate sustainability
  CSR, 102–105
  introduction, 101–102
  sustainability reporting for,
    102–108
Corruption
  control, 300t–302t
  human rights and, 254–256
  press freedom and, 253–254, 255f
Creativity
  aids, 72–81
  backcasting, 76
  brainstorming, 73–74
  case-based reasoning, inductive
    reasoning, and analogy,
    75–76
  ideas and, 73
  introduction to, 72–73
  mind-mapping, 74
  mood boards, 74
  from nature, 74–75
  9-windows method, 76–79, 78f
  post-normal science, 79–80, 79f
  reflection and, 73
  SCAMPER acronym, 75
  sketching, 74
  TRIZ analogy, 76–77

Critical elements. *See* Critical
    materials
Critical flags, 96
Critical materials. *See also specific*
    materials
  in aircrafts, 13, 15f
  in cell phones, 245, 245t
  classifying, 11
  description of, 89
  geopolitical risk of, 89–90
  history of, 11–13
  list of, 11
  origins of, 90f
  in telephones, 13, 14f
CSR. *See* Corporate social
    responsibility
Cycling
  driving compared to, 52
  electric bicycles, 166

D

Data, charts, and databases. *See*
    also Environmental data;
    Useful numbers
  CES EduPack sustainability
    database, 242–244
  elements data-table, 244–246
  energy storage systems
    data-table, 249–250
  introduction to, 242
  legislation and regulations
    data-table, 250–252
  materials data-table, 246–247
  nations of the world data-table,
    252–257
  power systems data-table,
    247–249
  summary of, 257–258
Data-intensity, 234t
Data-objects, 234t
Design
  circularity metrics, 230–231,
    230–231, 231f
  engineering, 31
  for environment, 217f, 218
  improvements to, 223–231,
    224f
  longer product life, 223, 225
  recovery obstacles, 229–230, 229f
  for recycling, 225, 228
  re-manufacturing, 227,
    227b–228b
  for repair, 225, 227
  for reuse, 220–221, 225–226
  waste in, 212

Developed nations, 28–29
Digital Lumens, 232b
Diverse materials
  abundance and price of,
    284t–285t
  environmental data for, 296t
Dodd-Frank act, 95–96
Dow Jones Sustainability World
    Index, 104
Driving. *See also* Vehicles
  carbon emissions and, 52
  cycling compared to, 52
  traffic lights, 274
  vehicle recycling, 83
Duty-cycle, 62

E

Earth elements. *See also* Rare earth
    elements
  abundance of, 95f
Eco-audits, 62–63
  in material data-table, 246–247
  for washing machine, 248f
Ecological footprint, 241f
Ecological metaphor, 213–217
Ecology, 235–236
  industrial, 217f, 218
Economic history, 2–3
Economic risk, 258
Economics
  bamboo flooring relating to,
    206
  biopolymers and, 126–127, 131
  electric cars relating to, 160
  fact-finding for, 64, 64b
  GDP per capita and price
    of energy and water,
    297t–299t
  lighting relating to, 176
  solar PV relating to, 190–191
  useful numbers for, 297–300
Economy. *See also* Circular
    materials economy
  capital relating to, 32, 70
  carbon taxes relating to, 33–34
  linear materials, 212, 219,
    234–235
  Western, 2–3
  wind farms relating to,
    144–145
Eco-sphere, 213–214
Edison, Thomas, 167–168
Efficiency factors, 291t–292t
Electric bicycles, 166
Electric buses, 165

Electric cars
  acceptance of, 154
  automobile associations and
    public on, 155
  background on, 152
  batteries for, 155, 157–158, 160
  carbon footprint of, 158–159
  case study on, 151–166
  cobalt in, 166
  concerns, 155b
  economics relating to, 160
  energy in, 157–158, 158f
  on environment, 158–159
  fact-finding on, 155–161, 156f
  government on, 154
  grants for, 153
  green campaigners on, 155
  human capital relating to, 163
  IC cars compared to, 151–152
  introduction, 151–152
  labor unions on, 155
  legislation on, 159–160
  lithium in, 156b, 157
  long-term analysis, 164
  manufactured capital relating
    to, 163
  materials in, 155–157, 156b,
    161, 166
  materials supply-chain for, 155
  natural capital relating to, 161,
    163
  in New York, 153
  Obama on, 153
  oil companies on, 155
  prime objective, 153
  projects relating to, 165–166
  reflection on, 163–164
  short-term analysis, 163–164
  society relating to, 160
  stakeholders of, 153–155, 154f
  synthesis, with three capitals,
    162t, 161–163
Electric hand dryers, 273
Electrical power generation,
  248–249
Electricity. *See also specific sources*
  generation, 286t
  global consumption of, 182
Elements
  abundance and price of, 246f,
    281t–282t
  data-table, 244–246
  diversity of, 9t
  environmental data for,
    293t–294t
  life, 214–215

ELV. *See* End-of-Life Vehicles
Embodied energy, 246, 247f
End-of-Life Vehicles (ELV), 251,
  251f
Energy
  for bamboo floors, 204–205,
    205f
  in biopolymers, 123–124, 125f
  conversion efficiency factors,
    291t–292t
  destinations, 17
  in electric cars, 157–158, 158f
  embodied, 246, 247f
  in Europe, 82
  fact-finding for, 62, 62b
  fuel intensity and carbon
    footprints, 62t, 285t
  GDP per capita and price of,
    297t–299t
  in Germany, 51–52
  global consumption of, 16f
  global production of, 17f
  of information, 234b
  intensity, 62t
  in lighting, 174
  lumens, global warming
    potential, and, 175t
  mix, 286t
  pay-back time, for wind, 142
  SI units, 279–280, 280t
  in solar PV, 187–188, 188f
  sources, 16, 16f
  of transport systems, 290t
  unit conversions for, 280t
  units and quantities, 16–17
  useful numbers for, 285–293
  vehicle energy ratings, 292t
  wind energy providers, 139
  in wind farms, 142
Energy storage
  grid-scale, 249
  importance of, 249
  in transport systems, 249–250
Energy storage systems
  data-table, 249–250
  performance metrics of, 289t
Engineering
  design, 31
  materials, 7
Environment
  bamboo flooring on, 205
  capital relating to, 70
  design for, 217f, 218
  electric cars on, 158–159
  fact-finding for, 62–63, 63b
  lighting on, 174

  solar PV on, 189
  technology relating to,
    217–219
  useful numbers for, 293–297
  wind farms on, 143
Environmental country risk, 257
Environmental data
  for diverse materials, 296t
  for elements, 293t–294t
  for polymers, 295t–296t
Environmental properties,
  125–126, 125f
Equilibrium, 216
Ethanol, 120
Ethical sourcing, 33–34
Europe, 82
European directives, 251
Extraction improvements, 222

F

Fact-finding, 266–267
  aid, 45f
  on articulation, 60–66
  on bamboo flooring, 202–207,
    203f
  on biopolymers, 122–127,
    123f
  in case studies, 112
  for economics, 64, 64b
  on electric cars, 155–161, 156f
  for energy and power, 62, 62b
  for environment, 62–63, 63b
  exercises on, 82–83
  in layered approach, 45,
    46b–47b, 48
  for legislation and regulation,
    63, 63b
  for lighting, 172–177, 172f
  for materials and manufacture,
    60–61
  six headings for, 61f
  for social equity, 63, 64b
  on solar PV, 186–191, 186f
  for time-value of money,
    64–65
  for wind farms, 140–145, 141f
Factual information, 40
Fairness, 70
Financial capital
  bamboo relating to, 208
  biopolymers relating to,
    129–130
  features of, 67t
  lighting on, 177
  solar PV relating to, 192
  wind farms relating to, 147

Flux, 168–169
Fossil hydrocarbons, 129
Freedom
    of press, 253–254, 255f
    of speech, 300t–302t
FTSE4Good Index, 104
Fuel
    biofuels, 33–34, 132–133
    CAFE rules, 252
    carbon footprint and intensity,
        62t, 285t
    intensity and carbon footprints,
        285t

G

Gas
    alternatives, 286t
    carbon footprint from, 143f
GDP. *See* Gross domestic product
Generalizations, 34
Geopolitical risk
    anticipating, 257
    of critical materials, 89–90
    HHI relating to, 89–90, 91f
    in materials supply-chain,
        89–91
Geothermal power, 273–274
Geranium, 99
Germany
    energy in, 51–52
    solar PV in, 182, 195
    wind farms in, 148–149
Global affluence, 253
Global Compact, 95–96
    UN, 103, 104b–105b
Global consumption
    of electricity, 182
    of energy, 16f
    of materials, 9–10, 219–220
    trade growth and, 17–18, 18f
Global financial resources,
        256–257
Global production
    of bamboo, 198, 204t
    of biopolymers, 119–120,
        131
    of energy, 17f
    of lithium, 157
    of materials, 10f
    of neodymium, 142, 155–156
    of rare earth elements,
        155–156
    of silicon, 187t
    of vehicles, 151
Global Reporting Initiative
        (GRI), 103

Global resource intensities,
        287t–288t
Global warming potential, 175t
Globalization, 94
Government. *See also* Legislation
    on electric cars, 154
    on wind farms, 139
Grants, 153
Green campaigners, 155
GRI. *See* Global Reporting
        Initiative
Grid-scale energy storage, 249
Gross domestic product (GDP),
        241f, 253
    life expectancy and, 255f
    price of water, energy, and,
        297t–299t
Groundnut Scheme, 114b
Growth
    of industrial systems, 216
    population, 17–18
    trade, 17–18, 18f
    well-being and, 216

H

Hand dryers, 273
Healthcare, 253
Herfindahl-Hirschman Index
        (HHI)
    exercises for, 98
    geopolitical risk relating to,
        89–90, 91f
History
    of critical materials, 11–13
    economic, 2–3
    of element diversity, 9t
    of materials, 7–10
    of natural fibers, 10
    of polymers, 10
    of sustainable development,
        3–7
Human capital, 31–32
    articulation relating to, 68b
    bamboo relating to, 208
    biopolymers relating to, 130
    electric cars relating to, 163
    features of, 67t
    lighting on, 177
    solar PV relating to, 192
    wind farms relating to, 147
Human development, 254f
    wealth and, 253
Human rights, 254–256
Hydrocarbons, 118
    fossil, 129
Hydrological cycle, 214–215

I

IC cars. *See* Internal combustion cars
Improved materials extraction
        and yield, 222
Improved materials technology,
        222–223, 224f
Incandescent lights. *See* Lighting
Incommensurate quantities, 40
India, 22
Indices. *See also* Herfindahl-
        Hirschman Index
    Community's CR Index, 104
    Dow Jones Sustainability World
        Index, 104
    FTSE4Good Index, 104
    SR, 104
Indium sourcing, 244
Individual learning views, 270
Inductive reasoning, 75–76
Industrial ecology, 217f, 218
Industrial lighting, 232b
Industrial revolution, 219
Industrial systems
    in eco-sphere, 213–214
    growth in, 216
    natural systems compared to,
        213–214, 216t
    resources in, 213
    temporal and spatial scale of
        thinking for, 217f
    waste in, 213
Information
    on biopolymers, 130
    carbon footprint of, 234b
    energy of, 234b
    factual, 40
    public, 130
    replacing materials by, 233
Informed synthesis, 66–70
Instructor guidance
    assessment rubric, 267–268,
        269t
    individual learning views, 270
    introduction to, 260
    learning outcomes, 263t–264t
    learning portfolios, 271
    PBL, 260–261
    project organization and
        scheduling activities,
        262–267, 263t–265t
    project suggestions, 272–274
    student aids, 270
    student feedback, 268–271
    summary, 271–272
    sustainable development and
        PBL, 261–262

Interest, 65–66
Internal combustion (IC) cars, 151–152
IPAT equation, 17–19

J

Jaguar Land Rover, 107b–108b

L

Labor unions, 155
Lamp bulbs. *See* Light bulbs
Landfill tax, 252
Landmark publications, 4, 5t
Layered approach
  assembling layers, 50, 51f
  fact-finding in, 45, 46b–47b, 48
  problem definition, 42–43, 42b
  reflection in, 49–50, 50b
  stakeholder identification, 43–45
  to sustainable development assessment, 42–50
  synthesis in, 48–49, 49b
Layering, 39f, 40, 41f
LCA. *See* Life cycle assessment
Learning
  individual views, 270
  outcomes, 263t–264t
  PBL, 260–262
  portfolios, 271
LEDs. *See* Lighting
Legislation
  on bamboo flooring, 206, 207t
  on biopolymers, 126
  CAFE rules, 252
  circular materials economy, 233
  Dodd-Frank act, 95–96
  on electric cars, 159–160
  ELV, 251, 251f
  European directives, 251
  examples of, 93t
  fact-finding for, 63, 63b
  globalization relating to, 94
  landfill tax, 252
  on lighting, 167–168, 174–176, 175t
  Montreal Protocol, 4
  on packaging, 126
  REACH Directive, 251–252
  by Rio Earth Summit, 4
  risk, 92–94
  on solar PV, 190, 190t
  WEEE Directive, 251
  on wind farms, 143–144
  by WSSD, 4

Legislation and regulation data-table, 250–252
Life cycle assessment (LCA), 62–63
Life elements, 214–215
Life expectancy, 300t–302t
  GDP and, 255f
Lifestyle changes, 233–234
  doing with less and doing without, 234
Light bulbs, 11. *See also* Lighting
Light emitting diodes (LEDs); Lighting
Light sources, 169t
Lighting
  background on, 168–169
  CFL lifespan, 174
  concerns, 171b
  Digital Lumens, 232b
  economics relating to, 176
  Edison and, 167–168
  energy in, 174
  on environment, 174
  fact-finding for, 172–177, 172f
  global warming potential, energy, and lumens, 175t
  on human and social capital, 177
  incandescent bulb bans, 169
  industrial, 232b
  introduction to, 167–169
  legislation on, 167–168, 174–176, 175t
  long-term analysis, 179
  lumen cost, 176t
  on manufactured and financial capital, 177
  materials in, 172, 173t, 174
  materials supply-chain in, 173–174
  mercury in, 174
  on natural capital, 177
  prime objective of, 169
  projects suggested for, 179–180
  reflection on, 179
  short-term analysis, 179
  society relating to, 176
  stakeholders in, 170–171
  synthesis, with three capitals, 177, 178t
  for traffic lights, 274
  unit measurements, 168–169
Linear materials economy, 212, 219, 234–235
Literacy, 300t–302t

Lithium
  in electric cars, 156b, 157
  global production of, 157
  nations producing, 157t
lm. See Lumen
Longer product life, 223, 225
Long-term sustainable development, 217f, 218
Lumen (*lm*), 168–169
  cost to provide, 176t
  energy, global warming potential, and, 175t
Lux (*lx*), 168–169

M

Magnets, 137f, 138t, 140–142, 141f
Malthus, Thomas, 3–4
Manganese, 99
Manufactured capital, 31–32
  articulation relating to, 68b
  bamboo relating to, 208
  biopolymers relating to, 129–130
  electric cars relating to, 163
  features of, 67t
  lighting on, 177
  solar PV relating to, 192
  wind farms relating to, 147
Manufacturing
  bamboo flooring, 199
  fact-finding for, 60–61
  re-manufacturing, 227, 227b–228b
Material criticality risk, 257
Material sourcing and usage. *See also specific materials*
  background on, 86
  constraint on, 86–88, 88f
Materials. *See also* Circular materials economy; Elements; *specific materials*
  in bamboo flooring, 202–203
  BOMs, 60–61
  chemical background of, 93–94
  conflict material risk, 257
  critical, 11–13
  data-table, 246–247
  diverse, 284t–285t, 296t
  in electric cars, 155–157, 156b, 161, 166
  element diversity, 9t
  embodied energy of, 246, 247f
  engineer, 97
  engineering, 7

exercises relating to, 20–23
expansion, 223b
fact-finding for, 60–61
global consumption of, 9–10,
    219–220
global production of, 10f
history of, 7–10
improved materials extraction
    and yield, 222
improved materials technology,
    222–223, 224f
before industrial revolution, 219
in industrial revolution, before
    and after, 219
information replacing, 233
in lighting, 172, 173t, 174
in linear materials economy,
    212, 219, 234–235
price movement of, 86–87, 87f
raw, 33–34, 86
in solar PV, 186–187
summary of, 20
in television, 21t
timeline, 8f
useful numbers for,
    280–285
in wind turbines, 136
Materials supply-chain, 61b
for electric cars, 155
in lighting, 173–174
Materials supply-chain risk
abundance risk, 94–95
conflict risk, 91–92
corporate responsibility in,
    95–96
exercises on, 97–99
geopolitical risk and supply
    monopoly, 89–91
introduction, 85–86
legislation and regulation risk,
    92–94
material sourcing and usage,
    86–88, 88f
price volatility risk, 88, 257
for raw materials, 86
risk management for, 96–97
summary of, 97
Matrix
stakeholder, 58–59, 59f
synthesis, 68–69, 69t
Mazuma Mobile, 226b
Mercury, 174
Michelin, 232b
Military spend, 256–257, 256f
Mind-mapping, 74

Mobile phones. *See* Cell phones
Modeling recovery cost, 230b
Money. *See also* Pay-back time;
    Price volatility
present value factor, 65
price movement, 86–87, 87f
recovery cost, 229–230, 229f,
    230b
time-value of, 64–65
Monopoly, 89–91
Montreal Protocol, 4
Mood boards, 74

N

Nations of the world data-table,
    252–257
Natural capital, 31–32
articulation relating to, 68b
bamboo relating to, 208
biopolymers relating to, 129
electric cars relating to,
    161, 163
features of, 67t
lighting on, 177
solar PV relating to, 191–192
wealth and, 253
wind farms relating to, 145–147
Natural fibers, 10
Natural systems
in eco-sphere, 213–214
industrial systems compared to,
    213–214, 216t
resources in, 213
temporal and spatial scale of
    thinking for, 217f
waste in, 213
Nature, 74–75
Neodymium, 136
future demand for, 161
global production of, 142,
    155–156
production, 142
Net comprehensive capital, 32
New York, 153
9-windows method, 76–79, 78f
Nitrogen cycle, 214–215
Nonstructural vehicle parts, 132

O

Obama, Barack, 153
Oceans, 22
Oil
alternatives, 286t
companies, 155
Oil-based polymers, 119t

P

Packaging, 126
Paper towels, 273
Pay-back time
of interest, 65–66
for solar PV, 190–191
for wind farms, 142
PBL. *See* Problem-based learning
PC and P. *See* Pollution control
    and prevention
Performance metrics, 289t
Personal satisfaction, 130
Physical constants, 277–280,
    277t–278t
Plastic bags, 82
Plastic bottles, 133
Platinum, 99
supply-chain, 13b
Political instability, 92
Pollution, 129–130
Pollution control and prevention
    (PC and P), 217–218, 217f
Polymers, 93–94. *See also*
    Biopolymers
abundance and price of,
    283t–284t
commodity, 10
environmental data for,
    295t–296t
history of, 10
oil-based, 119t
pollution from, 129–130
Population
affluence relating to, 17–19
in China, 22–23
exercises relating to, 20–23
in India, 22
IPAT equation, 17–19
present growth rates, 17–18
resources relating to, 3–4, 17–19
summary of, 20
of Uganda, 23
Post-normal science, 79–80, 79f
Power. *See also specific types of power*
fact-finding for, 62, 62b
SI units, 279–280, 280t
unit conversions for, 280t
Power systems
area-intensity of, 144f
capacity factors for, 287t–288t
data-table, 247–249
global resource intensities for,
    287t–288t
Present value factor, 65
Press freedom, 253–254, 255f

Pressure, 279, 279t
Price movement, 86–87, 87f
Price volatility
  of cobalt, 88, 89f
  of copper, 88, 89f
  of rare earth elements, 88, 89f
  risk, 88, 257
Prime objective
  of articulations, 33–34, 56–57
  of bamboo flooring, 200
  of biopolymers, 120, 130
  of case studies, 112
  clarification, 56–57
  electric cars, 153
  exercises on, 82
  framing, 266
  of lighting, 169
  size-scale, 56, 57t
  of solar PV, 183–184
  time allocation, 56, 57t
  of wind farms, 138
Problem definition, 42–43, 42b
Problem solving, 79–80, 79f
Problem-based learning (PBL),
    260–261
  sustainable development and,
    261–262
Product life
  longer, 223, 225
  of smart phone, 226b
  of wind turbines, 144–145
Production. *See also* Global
    production
  consumption compared to, 222
  neodymium, 142
Products
  services replacing, 231
  value of, 226f
Public information, 130
Publications, 4, 5t
PV panels. *See* Solar photovoltaic

R

Rare earth elements
  in cell phones, 272–273
  global production of, 155–156
  nations producing, 142t
  price volatility of, 88, 89f
Raw materials. *See also* Materials
  ethical sourcing of, 33–34
  materials supply-chain risk for,
    86
REACH Directive, 251–252
Rechargeable batteries, 93–94

Recovery carbon dioxide, 206
Recovery cost, 229–230, 229f,
    230b
Recovery obstacles, 229–230, 229f
Recycling
  design for, 225, 228
  recovery obstacles, 229–230,
    229f
  of vehicles, 83
Reflection, 71
  on bamboo flooring, 208–210
  on biopolymers, 130–132
  of case studies, 113
  creativity and, 73
  on electric cars, 163–164
  in layered approach, 49–50, 50b
  on lighting, 179
  on solar PV, 192–194
  by students, 267
  on wind farms, 147–148
Regulation. *See also* Legislation
  circular materials economy, 233
  fact-finding for, 63, 63b
  legislation and regulation
    data-table, 250–252
  risk, 92–94
Regulatory risk, 258
Remanent magnetization, 141f
Re-manufacturing, 227,
    227b–228b
Renault, 227b
Repair
  design for, 225, 227
  recovery obstacles, 229–230,
    229f
Reporting, 68–69. *See also* Sustain-
    ability reporting
Reports. *See also* Sustainability
    reporting
  Brundtland Report, 4, 6f
  on sustainable development, 4, 5t
Resources
  consumption of, 17–19
  global financial, 256–257
  global resource intensities,
    287t–288t
  in industrial system, 213
  in natural system, 213
  population relating to, 3–4,
    17–19
Reuse
  design for, 220–221, 225–226
  recovery obstacles, 229–230,
    229f

Rhodium, 99
Ricoh, 228b–229b
Rio Earth Summit, legislation by, 4
Risk. *See specific risk*
Risk management
  for abundance risk, 94–95, 95f,
    257
  critical flags of, 96
  for economic risk, 258
  for material criticality risk, 257
  for materials supply-chain risk,
    96–97
  for regulatory risk, 258
  risk anticipation, 257–258
  for social risk, 258
Rogers, Heather, 114b–115b

S

SCAMPER acronym, 75
Science-informed empiricism,
    222–223
Services, 231
SI units
  energy, 279–280, 280t
  physical constants in, 277–280,
    277t–278t
  power, 279–280, 280t
  stress and pressure, 279, 279t
Silicon, 186–187
  global production of, 187t
Silver, 187
Size-scale, 56, 57t
Sketching, 74
Smart phone, 226b. *See also* Cell
    phones
Social capital
  bamboo relating to, 208
  biopolymers relating to, 130
  features of, 67t
  lighting on, 177
  solar PV relating to, 192
  wind farms relating to, 147
Social equity, 63, 64b
Social responsibility. *See* Corporate
    social responsibility
Social risk, 258
Society
  bamboo flooring relating to,
    206–207
  corruption control, 300t–302t
  electric cars relating to, 160
  freedom, of press, 253–254,
    255f
  freedom, of speech, 300t–302t

human rights and corruption, 254–256
life expectancy, 255f, 300t–302t
lifestyle changes, 233–234
lighting relating to, 176
literacy, 300t–302t
solar PV relating to, 191
useful numbers for, 300–303
wind farms relating to, 140, 144
Solar chargers, 53
Solar photovoltaic (PV)
  background on, 182–183
  capacity factor, 183
  carbon footprint of, 186–187, 188f, 189
  concerns, 185b
  economics relating to, 190–191
  energy in, 187–188, 188f
  on environment, 189
  exercises on, 83
  fact-finding on, 186–191, 186f
  in Germany, 182, 195
  human and social capital relating to, 192
  industry growth, 183
  intermittency of, 183–184
  introduction to, 182–183
  irradiance global map, 182f
  legislation on, 190, 190t
  long-term analysis, 194
  manufactured and financial capital relating to, 192
  materials in, 186–187
  natural capital relating to, 191–192
  pay-back time for, 190–191
  prime objective of, 183–184
  projects suggested for, 194–195
  reflection on, 192–194
  short-term analysis, 192–194
  society relating to, 191
  stakeholders in, 184–185, 185f
  synthesis, with three capitals, 191–192, 193t
Sourcing risk, 257
SR. *See* Sustainability reporting (SR)
Stakeholder identification, 58, 266
  in layered approach, 43–45
Stakeholder interest, 58–59
  components of, 59f
Stakeholder matrix, 58–59, 59f
Stakeholder questions, 57–58
Stakeholders

analysis of, 57–60
in articulations, 34, 43–45, 57–60, 68b
bamboo flooring, 200–202, 201f
in biopolymers, 120–122
in case studies, 112
of electric cars, 153–155, 154f
exercises on, 82
influence of, 44, 44f, 57–59, 60f
in lighting, 170–171
in solar PV, 184–185, 185f
in wind farms, 138–140
Stress, 279, 279t
Strong sustainability, 32
Strontium ferrite, 82–83
Student aids, 270
Student feedback, 268–271
Student reflection, 267
Supply, 89–91. *See also* Materials supply-chain
Sustainability
  of coffins, 52–53
  corporate, 101–108
  defining, 28–29
  factors of, 29, 29t
  four principles of, 77t
  strong, 32
  of transportation, 28–29
  weak, 32
Sustainability reporting (SR)
  AngloAmerican case study, 106b–107b
  case studies, 105–108
  company performance and, 104
  for corporate sustainability, 102–108
  criteria, 105
  defining, 103
  GRI on, 103
  indices, 104
  Jaguar Land Rover case study, 107b–108b
  3BL reporting, 30–31
Sustainable business, 30–31
Sustainable development
  accountancy relating to, 30–31
  articulations of, 33–36
  Brundtland Report on, 4, 6f
  in business community, 6
  defining, 2–3, 30–32
  for engineering design, 31
  exercises on, 37
  history of, 3–7
  introduction, 2–3, 28

landmark publications on, 4, 5t
long-term, 217f, 218
operating principles, 218b–219b
PBL and, 261–262
reports on, 4, 5t
summary of, 36
Western nations on, 2
Sustainable development assessment
  complex systems, 40–41
  exercises, 51–53
  introduction, 39–40
  layered approach to, 42–50
  steps, 39–54
  summary of, 51
Synthesis
  bamboo flooring, with three capitals, 207–208, 209t
  of biopolymers, 127–130, 128t
  case studies, 112–113
  of electric cars, with three capitals, 161–163, 162t
  goal of, 66
  informed, 66–70
  in layered approach, 48–49, 49b
  of lighting, with three capitals, 177, 178t
  matrix, 68–69, 69t
  reporting of, 68–69
  solar PV, with three capitals, 191–192, 193t
  three capitals relating to, 66, 267
  of wind farms, with three capitals, 145–147, 146t

T
Taxes
  carbon, 33–34
  landfill, 252
Technological development, 3–4
Technology. *See also* Cell phones
  environment relating to, 217–219
  improved materials, 222–223, 224f
  lock-in, 114b
Telephones, 13, 14f. *See also* Cell phones
Television, 21t
Three capitals. *See also specific capitals*
  bamboo flooring relating to, 207–208, 209t

Three capitals. *See also specific capitals* (*Continued*)
  biopolymers relating to, 127–130, 128t
  defining, 31–32, 31f
  electric cars relating to, 162t, 161–163
  forces driving, 32
  impact of, 70f
  key features of, 67t
  lighting relating to, 177, 178t
  solar PV relating to, 191–192, 193t
  synthesis relating to, 66, 267
  in wind farms, 145–147, 146t
3BL reporting, 30–31
3Ps, of capital, 28
Time-value, 64–65
Trade growth, 17–18, 18f
Traffic lights, 274
Transport systems
  carbon footprint of, 290t
  energy of, 290t
  energy storage in, 249–250
Transportation
  in developed nations, 28–29
  driving compared to cycling, 52
  sustainability of, 28–29
TRIZ analogy, 76–77
Tungsten, 11

U
Uganda, 23
UN Global Compact. *See* United Nations Global Compact
Uncertainty, 79f
Unit conversions, 277–280, 278t–279t
  for energy, 280t
  for power, 280t
  for stress and pressure, 279, 279t
United Nations (UN) Global Compact, 103, 104b–105b
Useful numbers
  for economics, 297–300
  for energy, 285–293
  for environment, 293–297
  introduction to, 277–280
  for materials, 280–285

SI units, energy and power, 279–280, 280t
SI units, of stress and pressure, 279, 279t
SI units, physical constants, 277–280, 277t–278t
  for society, 300–303
  unit conversions, 277–280, 278t–279t

V
Value
  capital base, 48
  present value factor, 65
  of products, 226f
  time-, 64–65
Vehicles
  biocomposites, as nonstructural vehicle parts, 132
  CAFE rules, 252
  carbon dioxide ratings for, 292t
  carbon emissions and driving, 52
  driving compared to cycling, 52
  electric bicycles, 166
  electric buses, 165
  electric cars, 151–166
  energy ratings for, 292t
  global production of, 151
  IC cars, 151–152
  Michelin, 232b
  recycling of, 83
  Renault, 227b

W
Washing machine, 248f
Waste
  in design, 212
  in industrial system, 213
  in natural system, 213
Water
  in biopolymers, 125f
  GDP per capita and price of, 297t–299t
  hydrological cycle, 214–215
Weak sustainability, 32
Wealth
  healthcare relating to, 253
  human development and, 253
  natural capital and, 253
WEEE Directive, 251

Well-being, 216
Western nations, 2–3
Wind farms
  background on, 136–137
  communities relating to, 140, 144
  concerns, 140b
  economy relating to, 144–145
  energy in, 142
  energy pay-back time for, 142
  energy providers for, 139
  on environment, 143
  fact-finding for, 140–145, 141f
  in Germany, 148–149
  government on, 139
  human and social capital relating to, 147
  introduction, 135–137
  legislation on, 143–144
  long-term analysis, 148
  manufactured and financial capital relating to, 147
  natural capital relating to, 145–147
  obligations of, 149
  prime objective of, 138
  projects relating to, 148–149
  reflection on, 147–148
  short-term analysis, 147–148
  society relating to, 140, 144
  stakeholders in, 138–140
  synthesis, with three capitals, 145–147, 146t
Wind speeds, 137f
Wind turbines, 56
  building rates, 142
  exercises on, 84
  life of, 144–145
  magnets, 137f, 138t, 140–142, 141f
  makers of, 140
  materials in, 136
World Business Council for Sustainable Development, 6
World Summit on Sustainable Development (WSSD), 4

Y
Yield improvements, 222
Young's modulus, 123, 124f
Yttrium, 174

Printed in the United States
By Bookmasters